Who Can You Trust?

'In a time when people are doubting experts, suspicious of the media, and losing faith in government and business, Rachel Botsman is here with a lucid analysis of what it takes to build and rebuild trust. Trust me: this is a book you need to read' Adam Grant, *New York Times* bestselling author of *Give and Take, Originals* and *Option B* with Sheryl Sandberg

'Rachel Botsman's eye-opening, timely book delves into the unfolding crisis of trust spreading throughout the world. She brilliantly describes how the established trust framework is undergoing a radical transformation as digital technologies take root in every facet of our lives. Read this book and you'll be ready for a revolution in trust that rewrites the rules of human interaction' Marc Benioff, Chairman & CEO, Salesforce

'In this extremely thought-provoking new book, Rachel Botsman educates and entertains as she reveals with expertise how our lives are already changing more than we know. A must-read for anyone interested in how the world works – and will work in the future' Will Dean, MBE, CEO, Tough Mudder

'Not only is the thesis completely compelling. It demonstrates, through a sequence of real-world case studies of projects, businesses and platforms, that the distributed trust model offers enormous promise if used wisely – as well as enormous pitfalls if used unwisely. For good and sometimes ill, it has the potential to reshape everything we do, from our choice of babysitter to our choice of money. These are important messages from what is an important book' Andy Haldane, Chief Economist, Bank of England

'A fascinating and well-researched study. Every reader will gain new insights into one of the great issues of our time: the shifting tides of trust' Geoff Mulgan, CBE, Chief Executive of the National Endowment for Science Technology and the Arts (Nesta)

'Profound insights about how the digital age changes trust, wrapped in a compelling narrative of captivating and revealing stories. A rare book that will cause you to think deeply about your business, your relationships and your life' Don Tapscott, bestselling author of *Wikinomics* and *Blockchain Revolution*

'Thi … resent and future of t… s unmatched. It's an a… consumers alike' Nic… nt, Airbnb

'This is that admirable and all-too-rare book that gives you "an idea to think with" that helps to put new things in place. *Who Can You Trust?* is a primer for a new world that sets you up to be a better citizen, consumer and parent. I quickly learned so much about so many things I wanted to know' Professor Sherry Turkle, author of *Reclaiming Conversation* and *Alone Together*

'This is a book that every adult reader should pick up to gain some perspective on how reliant we have become on technology, and how we could afford to approach it with a little more skepticism' *Library Journal Review*

'Botsman has found a rich theme here and a fascinating way of interpreting the technological change' *Wall Street Journal*

'A timely and accessible framework for understanding what trust is, how it works, why it matters and how it is evolving' *Washington Post*

'A well-researched and insightful book, which explores a vital cultural and social reality. Botsman rightly challenges us in this new era to ask the compelling questions about who, why and how we trust. Highly recommended' Tim Costello, CEO, World Vision Australia

'Some people can educate and others can entertain; in *Who Can You Trust?* Rachel does both as she gives us her unique insight into the need for and the leverage of trust, Read it for insight or escape as it takes you on both journeys' John Eales, most successful captain in the history of Australian rugby

'Trust hasn't shrunk, it's shifted – and it's a shift that's changing the world. In this compelling account, Rachel Botsman captures all the chaos and challenge; all the danger and opportunity. Timely, lucid and beautifully written. This is one of the most important books you'll read this year' Richard Glover, Columnist, *Sydney Morning Herald*, and ABC radio broadcaster

'Like "time", "trust" is a phenomenon that we intuitively understand but find almost impossible to explain. However, in Rachel Botsman's capable hands, the concept of "trust" – and its changing shape over the ages – becomes clear and accessible. Rachel Botsman drives home a central point: we cannot "outsource" personal responsibility to bots, networks or systems. Managing trust in the digital age is a human challenge. Utterly compelling' Dr Simon Longstaff, Executive Director of the Ethics Centre

'Rachel Botsman's thrilling account of the onrush of information technology explains why Bill Gates, Stephen Hawking and Elon Musk all fear the next great leap in the development of artificial intelligence. Botsman brilliantly exposes the central paradox of the IT revolution – that it connects us while keeping us apart . . . she encourages us to take responsibility for the kind of world we want to live in, and to preserve society's most fragile asset: trust' Hugh Mackay, founder of The Ipsos Mackay Report

'Botsman guides the reader on an enjoyably accessible, but cautiously skeptical, tour through this hugely transformative, but barely recognized, shift in our sometimes-irrational approach to trust' *Winnipeg Free Press*

'As Botsman drives deeper and darker, she sheds more and more light, her book rapidly becoming brilliant . . . Botsman's closing suggestion of a "trust pause" – a period of cooling off before granting our trust – forms a decent starting point for answering one of the biggest questions of our time: Who can we trust? It would a good idea to find out, before others decide it for us' *Dialogue*

ABOUT THE AUTHOR

Rachel Botsman is a world-renowned expert on trust, whose three TED talks on the topic have been viewed over 4 million times. She is also a lecturer at the University of Oxford's Saïd Business School.

Rachel was named one of the world's top twenty speakers to keynote your conference by *Monocle*, one of the Most Creative People in Business by *Fast Company* and a Young Global Leader by the World Economic Forum. She writes for the *New York Times*, *Guardian*, *Wall Street Journal*, *Harvard Business Review*, *Wired* and more. In her first book, *What's Mine is Yours*, she predicted the rise of the sharing economy. The concept was subsequently named by *Time* as one of the 'Ten Ideas That Will Change the World'.

Who Can You Trust?

How Technology Brought Us Together – and Why It Could Drive Us Apart

RACHEL BOTSMAN

BUSINESS

PENGUIN BUSINESS

UK | USA | Canada | Ireland | Australia
India | New Zealand | South Africa

Penguin Business is part of the Penguin Random House group of companies
whose addresses can be found at global.penguinrandomhouse.com.

First published by Portfolio Penguin 2017
Updated edition published by Penguin Business 2018
001

Illustrations by Team (www.team.design)

Set in 10.68/13.20 pt Dante MT Std
Typeset by Jouve (UK), Milton Keynes
Printed and bound in Great Britain by Clays Ltd, Elcograf S.p.A.

A CIP catalogue record for this book is available from the British Library

ISBN: 978–0–241–29618–9

www.greenpenguin.co.uk

Penguin Random House is committed to a sustainable future for our business, our readers and our planet. This book is made from Forest Stewardship Council® certified paper.

In memory of Pamela Hartigan, my friend and mentor.

What is trust?

A confident relationship with the unknown.

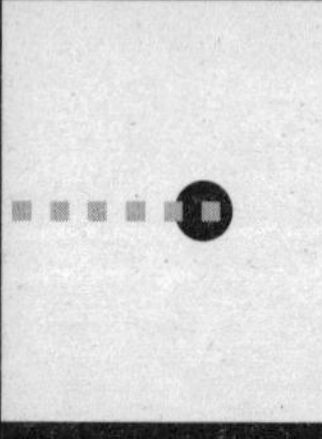

INSTITUTIONAL

LOCAL

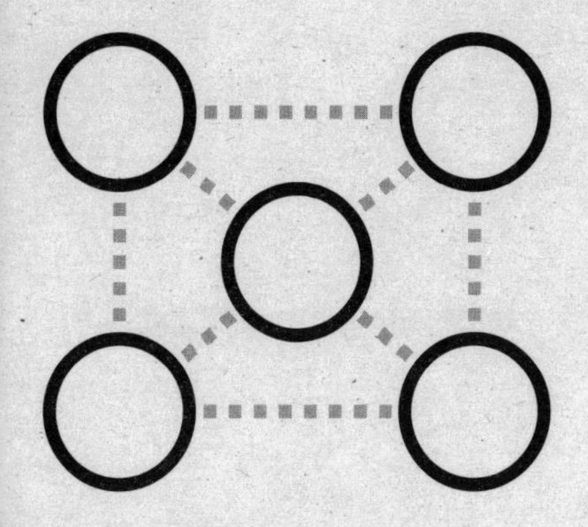

DISTRIBUTED

EVOLUTION OF TRUST

Contents

Introduction

'Abandon weapons first, then food. But never abandon trust. People cannot get on without trust. Trust is more important than life.'

Confucius to his disciple Tzu-Kung

I was getting married on the day the hammer fell on Wall Street. The date was 14 September 2008. I had been living in New York for almost a decade and had met my then fiancé, Chris, in a downtown dive bar called Eight Mile Creek. We were both 'city people' but we wanted to have our wedding in a rural, rustic setting. The place we finally chose was called Gedney Farm, nestled in the charming old Berkshire village of New Marlborough, Massachusetts.

'So, you want to get married in a horse barn?' my father said when I showed him the venue, a Normandy-style red barn surrounded by lush meadows and abundant orchards. Getting into the spirit of things after that, he decided we should arrive at the venue in an old-fashioned horse and carriage. I went along with his Cinderella fantasy and climbed into an open-topped white carriage, complete with a driver and a footman, drawn by an old grey mare. The horse, on its last legs, was slow. It rained. I was late.

Around eighty guests, our closest family and friends from all over the world, joined us for the occasion. Lit by candles and strings of Edison bulbs, the ceremony was traditional and very beautiful. The best man's speech was funny and the food was delicious, despite my finding a grasshopper about the size of my little finger in the green salad.

So there I was, at the heart of one ancient institution – marriage – built on trust and life-long commitment, while another – Wall Street – was imploding. Lost in the bubble of the celebrations, I didn't realize the outside world was in meltdown until around 9.30 p.m., when I finally noticed that, around the room, the warm glow of the Edison bulbs was competing with the brash blue glare of iPhones and BlackBerries as guests stealthily consulted their hand-held harbingers of doom. Family and friends who worked in banking were trying to absorb the barrage of messages flooding in. Could the impossible have happened? Lehman Brothers had just filed for Chapter 11 bankruptcy protection. Bank of America and Barclays had pulled out of a deal that might have saved the 158-year-old firm. Merrill Lynch had agreed to be bought by Bank of America for roughly $50 billion* in an attempt to avert a financial crisis. Washington Mutual, Wachovia and HBOS in the United Kingdom were within a whisker of collapsing. The fate of another giant, American International Group (AIG), the vanguard of the credit default swap market, teetered in the balance.

A couple of friends who were senior executives at JP Morgan Chase and Goldman Sachs apologized for having to leave, summoned to 'red alert' emergency meetings. It would be a race against the clock to avoid the blind panic that would surely happen when the markets opened. Several other guests drank nervously and partied hard, not sure if they would be carrying their work belongings out in boxes the following day. We danced the Horah, a traditional Jewish wedding ritual, which ended with me being elevated on a chair and my husband being thrown precariously up in the air on a large white tablecloth. Another moment of trust. Guests whirled around us, clapped and made 'Oy! Oy! Oy!' noises. Meanwhile, outside the barn, the biggest global financial crisis in history was building up a head of steam.

It was, of course, the beginning of the nerve-shattering period when many businesses 'fell off a cliff' and the world's financial system came closer to collapse than at any time since the Great Depression. As we now know, the economic repercussions of the meltdown would engulf the world for many years to come. But my

* 'Dollars' and the dollar symbol ($) refer to US dollars throughout.

wedding day, rich with tradition, also marked the downfall of something more profound: public trust in institutions.

Who was to blame for the crisis? What were the main causes? These questions were at the heart of the Financial Crisis Inquiry Commission (FCIC) created to investigate the banking collapse, and the answer was damning. 'The crisis was the result of human action and inaction, not of Mother Nature or computer models gone haywire,' the 525-page report found. 'To paraphrase Shakespeare, the fault lies not in the stars, but in us.'[1] In other words, the meltdown was an 'avoidable' human disaster.

The federal inquiry hammered the embarrassing failures of regulators, whom the report described as 'sentries not at their posts'. The finger was pointed squarely at the Federal Reserve for its failure to question widespread, egregious mortgage lending, over-reliance on short-term debt and the excessive packaging and reselling of loans, along with many other red flags. According to the report, however, the main culprit was not the toxic financial instruments but the human failings that drove them: reckless risk-taking, greed, incompetence, stupidity and a systemic breakdown in accountability and ethics.

It wasn't the first nail in the coffin of institutional trust and it probably won't be the last, but the financial crisis struck deep.

A loss of trust amounts to a lack of faith and confidence in 'the system' itself. What should we believe in if the system has failed us? Who, or what, can be relied upon? We begin to fear what else can go wrong. What other shortcomings we don't yet know about might lurk in the system? Fear, suspicion and disenchantment are deadly viruses that spread fast. The initial epicentre of the trust explosion was, understandably, with the banks. But it hasn't stopped there. Since the crisis, other scandals, other revelations, have seen the ripples of distrust touch government, the media, charities, big business and even religious organizations.

Like the plot of some overblown soap opera or Jacobean tragedy, the episodes of unethical behaviour have come thick and fast, from the lurid, even criminal, to the just plain stupid and, sadly, routine. Each has chipped away at public confidence. The British MPs' expenses scandal; the false intelligence about weapons of mass destruction (WMDs); Tesco's horsemeat outrage; price gouging by big pharma; the BP Deepwater Horizon oil spill; the dishonours of FIFA's bribery; Volkswagen's 'dieselgate'; major data breaches from companies such as Facebook and Cambridge Analytica, Sony, Target and Equifax; the Panama and Paradise Papers and widespread tax avoidance; the exchange-rate manipulation by the world's largest banks; Brazil's Petrobras oil scandal; the lack of an effective response to the refugee crises; and, last but not least, shocking revelations of widespread abuse by Catholic priests, other clergy and other 'care' institutions. No wonder a thousand headlines lament that nobody trusts authority any more. Corruption, elitism, economic disparity – and the feeble responses to all of the above – have pummelled traditional trust in the old institutions as fiercely as a brutal wind lashing ancient oaks.

Significantly, this crisis is taking place in a landscape of rapidly shifting and evolving technologies, from artificial intelligence (AI) to automation to the Internet of Things (IOT). We are already putting our faith in algorithms over humans in our daily lives, whether it's trusting Amazon's recommendations on what to read or Netflix's suggestions on what to watch. But this is just the beginning. We will soon be riding around in self-driving cars, trusting our very lives to the unseen hands of technology.

At the same time, many people are feeling so overwhelmed by the pace of change and the sheer amount of knowledge now available at a swipe or keystroke that they are beating a retreat to media echo chambers that narrow down information and reinforce already held beliefs. It becomes easy to ignore or simply not see contrary views. Technology, for all its pluses, also means falsehoods and so-called 'fake news' can quickly spread through networks unchecked and with an unstoppable momentum. Fake news has become a game of accusation and counter-accusation. If it started out as a useful identifier of misinformation, it is now an unhelpful catch-all term hurled at all kinds of uncomfortable truths a president, say, might not like. In fact, online misinformation on a grand scale – and the potential for digital wildfires – was listed by the World Economic Forum (WEF) in 2016 as one of the major risks to our society.[2] The result of those echo chambers and that misinformation? Our fears are verified, often baselessly. Our anger is amplified. The cycle of distrust is magnified. All in all, our faith in many institutions has been dragged to a critical tipping point.

Indeed, recent gloomy poll numbers would have any politician or business leader in a sweat. For the past seventeen years, the global communications firm Edelman has been conducting an annual 'Trust Barometer', asking more than 30,000 people across twenty-eight countries about their level of trust in various institutions. The headline for the 2017 results was, tellingly, 'Trust in Crisis'. Trust in all four major institutions – government, the media, business and non-governmental organizations (NGOs) – is at an all-time low.[3] The media suffered the biggest blow, now distrusted in 82 per cent of all countries surveyed. In the UK, the number of people saying they trusted the media fell from 36 per cent in 2016 to 24 per cent in 2017. 'People now view media as part of the elite,' says Richard Edelman, President and CEO of PR firm Edelman. 'The result is a proclivity for self-referential media and reliance on peers.'[4] In other words, looking to reinforce what we already believe, often from people we know.

The Brexit vote to leave the European Union and the election of Donald Trump are the first wave of acute symptoms emerging from one of the biggest *trust shifts* in history: from the monolithic to the individualized. Trust and influence now lie more with 'the people' – families, friends, fellow users, colleagues, even strangers – than with top-down elites, experts and authorities. It's an age where individuals can have more sway than traditional institutions and customers are not just meek consumers but social influencers that define brands.

By asking challenging questions about the flawed structure and size of institutional systems, and who runs them, we are coming to another confronting realization. Institutional trust, taken on faith, kept in the hands of a privileged minority and operating behind closed doors, simply wasn't designed for the digital age.

It wasn't designed for an age of radical transparency, of WikiLeaks and Cryptome, where politicians and CEOs must imagine they are operating behind clear glass. Trying to hide, well, anything really, is a high-stakes gamble. It doesn't work in a world where PR puffery can no longer cover up dirty secrets or closed-door antics. Take a few recent examples of 'private' matters that have been spilled around the world: the sensitive user data of the extra-marital dating site Ashley Madison, Turing Pharmaceuticals' internal emails on its predatory drug pricing, secret Scientology manuals, Hillary Clinton's emails and even a private conversation that took place within a private palace garden between the Queen of England and the Metropolitan Police Commander about the rudeness of Chinese officials.[5]

It wasn't designed for an age where people can transact directly on platforms such as Airbnb, Etsy and Alibaba. It wasn't designed for an era where it is predicted half of the workforce will be 'independent workers' – freelancers, contractors and temporary employees – within the next decade. It wasn't designed for a time where we have become dependent on tech powerhouses such as Facebook and Google which represent new forms of 'network monopolies' and platform capitalism. It wasn't designed for a culture where we want to control everything personally, from our bank accounts to our dates, with a swift click, tap or a swipe.

So should we be mourning the loss of trust? Yes, and no, because here's the thing: whatever the headlines say, this isn't the age of distrust – far from it. Trust, the glue that holds society together, hasn't disappeared. It has shifted – and the implications, for everything from hiring a babysitter to running a business, are massive.

For the past decade, I have been researching how technology is radically changing our attitudes towards trust. In 2008, I started writing my first book, *What's Mine is Yours*, about the so-called 'collaborative' or 'sharing economy'. I was fascinated by how technology could unlock the value of idle assets – cars, homes, power drills, skills, time – but it was the trust ingredient, how technology could make us engage in behaviours that might previously have been considered a little creepy or outright risky, that became my obsession.

Even then, the notion of building a marketplace based on letting strangers stay in other people's houses seemed ludicrous. Today, Airbnb, the home-sharing marketplace, is valued at $31 billion, making it the second most valuable hospitality brand in the world.[6] In 2008, it was hard to see how detailed online profiles would give people the confidence to get lifts with strangers operating as cab drivers and using their own cars. Today, Uber is valued at $72 billion, making it one of the biggest companies in the world, larger than FedEx, Deutsche Bank or Kraft Foods.[7] And then there is the explosion of online dating apps such as Tinder, where the average number of daily swipes is more than 1.4 billion with 26 million matches made daily.[8] These are just a handful of examples where online tools are enabling us to have face-to-face interactions and entrust strangers with our most valuable possessions, experiences, even our lives, in previously unimaginable ways.

Consider this: why do people say they don't trust bankers or politicians yet trust strangers to share a ride with them?

One conventional explanation is that people don't always tell the truth in surveys. That may be so, but there had to be more to this trust paradox. I had a hunch something deeper was happening. What if trust, like energy, cannot be destroyed and instead just changes form?

Who Can You Trust? charts a theory, a bold claim: we are at the start of the third, biggest trust revolution in the history of humankind. When we look at the past, we can see that trust falls into distinct chapters. The first was *local*, when we lived within the boundaries of small local communities where everyone knew everyone else. The second was *institutional*, a kind of intermediated trust that ran through a variety of contracts, courts and corporate brands, freeing commerce from local exchanges and creating the foundation necessary for an organized industrial society. And the third, still very much in its infancy, is *distributed*.

A trust shift need not mean the previous forms will be completely superseded; only that the new form will become more dominant. For example, a small farming community may continue to rely on centuries-old local trust in some matters, but turn more often to the new town court to handle others.

Trust that used to flow upwards to referees and regulators, to authorities and experts, to watchdogs and gatekeepers, is now flowing horizontally, in some instances to our fellow human beings and, in other cases, to programs and bots. Trust is being turned on its head. The old sources of power, expertise and authority no longer hold all the aces, or even the deck of cards. The consequences of that, good and bad, cannot be underestimated.

The explosive growth of the sharing economy is a textbook example of *distributed trust* at play. But the theory is also a way to understand the rapid evolution of platforms like the darknet, where consumers are happily scoring everything from marijuana to AK-47s from 'untrustworthy' dealers. The darknet and the new era of digitally enabled app intimacy may sound as if they have little in common, but they share the same underlying principle – people trusting other people through technology.

Distributed trust explains why we are now feverishly scoring and rating everything from restaurants to chatbots to Uber drivers (and why passengers are rated, too), helping to shape, almost instantly, the rise or fall of all sorts of businesses, while also creating *reputation trails* where one mistake or misdemeanour could follow us potentially for the rest of our lives.

Distributed trust helps us understand why digital cryptocurrencies such as bitcoin and ether could be the future of money, and how the blockchain (the underlying ledger technology that powers these cryptocurrencies) could be used for everything from tracking the source of foods or blood diamonds to selling our homes without the need for estate agents.

Distributed trust helps us grapple with why and how we'll come to trust well-trained bots, whether they're giving us relationship advice, resolving our parking tickets, ordering our sushi or telling us if we have cancer.

Indeed, I believe the real disruption happening is not technology itself, but the massive trust shift it creates.

Distributed trust is not simply a new, idealistic flavour of techno libertarianism. There are many stories in this book that show how it can have negative, dark or disastrous consequences – discrimination, theft and even death. Yes, technology can widen the circle of trust, unlocking the potential to collaborate and connect with unfamiliar strangers, but it can also erect and harden boundaries between us. Ratings and reviews may make us more accountable, even a little nicer, to our fellow human beings but our growing reliance on them also means some people will become forever tarnished, relegated to a kind of digital purgatory. And, in our rush to reject the old and embrace the new, we may end up placing too much trust, too easily, in the wrong places. One of the most pressing issues of our time is whether technology in fact helps us to make better or worse choices about where to place our trust.

It's already becoming clear that the turpitudes of institutions, real or fabricated, have left many people dangerously receptive to alternatives, and ready to place unquestioning faith in a new, and some would say highly dubious, breed of trust arbiter. Distributed trust is far from foolproof and the questions that really matter are ethical and moral, not technical.

The first two chapters of this book pose a simple question: how did we end up here? They unpack why trust matters so much. The next three chapters explore the trio of conditions that make distributed trust possible – trust in a new idea, trust in platforms and, finally, trust in other people or bots. This section explains how to adapt to building trust in this new era and what to do when it's lost. Critically, it asks who takes responsibility when trust is no longer centralized but distributed.

Elsewhere, the book travels to the depths of the darknet to understand why reputation matters so much, even to cocaine dealers. It goes inside the Orwellian-like trust-scoring system that is emerging in China and could determine everything from a citizen's job to whether they can get on a train or a plane.

The final chapters look to our digital future, particularly focusing on our rapidly evolving trust in artificial intelligence. If we make a habit of trusting intelligent machines, does it become harder to build trusting relationships with people? The glorified promises of the blockchain are explored. Will this digital ledger really become the 'Internet of Value', as many enthusiasts claim? Will the big banks end up 'taking over' this technology originally designed to cut out the middlemen?

Distributed trust, enabled by new technologies, is rewriting the rules of human relationships. It's changing the way we view the world and each other, returning us to the old village model of trust in one sense, except that the community is global in scale and some of its invisible reins are being pulled by internet giants. Now more than ever it is critical to understand the implications of this new trust era: who will benefit, who will lose and what the fallout might be.

Why? Because without trust, and without an understanding of how it is built, managed, lost and repaired, a society cannot survive, and it certainly cannot thrive. Trust is fundamental to almost every action, relationship and transaction. The emerging trust shift isn't simply the story of a dizzying upsurge in technology or the rise of new business models. It's a social and cultural revolution. It's about us. And it matters.

1
Trust Leaps

Friday, 19 September 2014, was a historic day on Wall Street. From the moment the markets opened at 9.30 a.m. Eastern Standard Time (EST), the ticker rocketed for one company in particular. It was called Alibaba. By the end of the day, the Chinese e-commerce giant had a staggering market capitalization of $231 billion.[1] It was the largest global public flotation ever on the New York Stock Exchange (NYSE), dwarfing that behemoth known as Facebook and even Alibaba's giant rival, Amazon. Overnight, fifty-year-old Chinese businessman Jack Ma, the company's founder and current chairman, became a very, very wealthy man.

A massive crowd lined the streets and packed on to the floor of the New York Stock Exchange that day to get a glimpse of the legendary entrepreneur. He was greeted like a rock star. 'What we raised today is not money, it's the trust from the people,' Ma told more than a thousand cheering admirers.[2]

It was not, however, the charismatic and dynamic founder who rang the famed opening bell at the Stock Exchange. Instead, Ma opted to have eight Alibaba customers, five of whom were women, to stand on the podium to start the day of trading. He wanted to show he was true to his mantra, 'Customers first, employees second and shareholders third.' One merchant – one of the millions of small businesses who trade on Alibaba's sites – was Lao Lishi, a former Chinese Olympic diving gold medallist who sells wooden beaded bracelets. Another was Peter Verbrugge, an American farmer who currently holds the record for selling the greatest amount of cherries through Alibaba.[3] The customers ringing the NYSE bell represented something very important to Ma – how

Alibaba has transformed the way Chinese businesses of all shapes and sizes can buy and sell a bewildering variety of goods, from clothes and nappies to live pedigree dairy goats and frozen chicken feet, inflatable sex dolls and even 'do-it-yourself abortion kits' to people all over the world.

But Jack Ma's story is not simply a fascinating rags-to-riches tale of entrepreneurial persistence. It also represents a remarkable feat in the delicate business of building trust.

It's a challenge to build any successful online marketplace where two sides need to trust one another, but what makes Ma's story extraordinary is that he achieved this in China. Traditionally, China is a society based on the concept of *guanxi*, loosely translated as 'relationships'. Trust, in business as well as private life, exists between people in the same *guanxi*: family, friends and people in the same village. People they know well over time, in other words, not strangers on a far-flung planet called the internet. In fact, it is common to *distrust* people outside your own personal network. This can create a cultural impediment and business obstacle as people are more prone to avoid building new relationships where there is no close connection.

I first went to Shanghai on business when I was twenty-five. I was part of a consulting project for a well-known brand looking to expand into Asia. Over the course of the first week, we shared an array of meals with our Chinese business clients. Lazy Susans spun, we ate delicious food at lunch and dinner, and we clinked beer glasses in toast after toast. The gatherings were warm and enjoyable but by day three I was wondering when we would get to the 'real' work. Rather insensitively, I didn't realize how important it was for Chinese businesspeople to spend a considerable amount of time socializing and getting to know you at the start of a relationship. 'In the West, we tend to reserve trust from the heart (affect-based trust) for family and friends and trust from the head (cognition-based trust) for business partners,' explains Professor Paul Ingram of Columbia Business School, who studies social networks. 'But in China, affect- and cognition-based trust are highly entwined even in business.'[4] It's especially true in China that people only trust you after you have invested a lot of time upfront in proving yourself to be *trustworthy.*

That was what Jack Ma was up against. It was one of the rock-hard conventions of trust he would set about shattering.

Ma Yun, as he was originally named, was raised in Hangzhou, about a hundred miles southwest of Shanghai, during Mao's Cultural Revolution. The middle child of three, his parents were traditional musical theatre performers. Ma inherited their love of showmanship. He would later get a reputation for dressing up in elaborate wigs and leather gear, and belting out theme songs from the *Lion King* at company events.

Ma wasn't a great student at school, but he was canny. From a young age, he realized the importance of mastering English. After President Nixon visited Hangzhou in 1972, tourists flooded to the area to see the beautiful lakes, temples and gardens. Every day, Ma would set his alarm early, around 5.00 a.m., and ride his bike to the Hangzhou Hotel. He would talk to the guests in English and offer to act as a guide, giving tours around the city free of charge. He did this for more than nine years. 'These Western tourists opened my mind because everything they told me was so different from the things I learned from school and from my parents,' Ma has said.[5]

Over the years, Ma befriended many tourists including a young American woman who suggested he take on an English name. Her husband's name was Jack, as was her father's. And so Jack he became.

Jack Ma would go on to become China's richest man in 2014, worth more than $19.5 billion, hauling himself to that lofty position with the aid of a shatter-proof resilience in the face of failure, a good dose of raw ambition, and another vital kind of trust – trust in himself.[6] He applied to Harvard ten times and was rejected on every occasion. (Who applies ten times?) He failed China's national university entrance exam twice. When he did eventually graduate in 1988 with a degree in English, he became a schoolteacher.[7] To supplement his modest $3 a week income, he would buy and sell plastic carpets on the streets of Hangzhou. At heart, he was a businessman.

With China's economy steadily improving and ideological barriers falling in the early 1990s, Ma decided to quit teaching. He applied for more than thirty different jobs but missed out on them all. When he applied to be a police officer, he was told in four simple words, 'You are no good.' 'I even went to Kentucky Fried Chicken when it came to China,' Ma told an audience at the World Economic Forum in Davos. 'Twenty-four people went for the job. Twenty-three were accepted. I was the only guy who wasn't.'[8]

It was in 1995, on his first visit to the United States, that his life took a fortunate turn. Ma had started Hangzhou Hope Translation Agency a year earlier, and went to America to help a Chinese firm sort out a financial dispute they were having with a US partner. It turned into a terrifying trip – the American he had been sent to see was a con man who flashed a gun at him. He made his way to Seattle to stay with Stuart Trusty, a friend who happened to run one of American's first internet providers, VBN.[9] Ma noticed a mysterious grey box with a screen sitting on his friend's desk. He wondered what on earth it was. 'Jack, it's not a bomb,' Trusty reassured him. 'It's a computer. Search for anything you want.'

Ma slowly typed in the word 'beer'. He doesn't remember why, possibly because it is easy to spell. A list of beers from Germany, America and Japan popped up, but noticeably no Chinese beers. Next, he typed in 'beer' and 'China'. No results. Keep in mind this was 1995. Netscape had just launched and Yahoo was an infant. Google would not launch for another three years. It was still the days of achingly slow dial-up internet. Nonetheless, Ma sensed something huge waiting in the wings.

Back home, he started China Pages, a kind of online Yellow Pages for Chinese companies. 'The day we got connected to the web, I invited friends and TV people over to my house . . . we waited three and a half hours and got half a page,' he says. 'We drank, watched TV and played cards, waiting. But I was so proud.

I proved [to my house guests that] the internet existed.'[10] Ma eventually sold the directory business to the state-run Hangzhou Telecom for 1 million yuan (approximately $148,000), a very large sum of money at the time. Next, he headed to Beijing to advise the Ministry of Foreign Trade and Economic Cooperation on ways to get Chinese companies to take up 'electronic commerce'.[11] 'My boss wanted to use the internet to control small businesses but I wanted to use the internet to empower small businesses,' he says.[12] What Ma really wanted to do was to start and build companies.

Ma's revolutionary vision was to help transform China's entire export economy, connecting small- to medium-sized Chinese businesses to Western customers and Western companies to a myriad of Chinese factories. Amazon and eBay had launched in the United States but nothing like this existed yet in China.

Today, more than 80 per cent of *all* goods bought and sold online in China are through Alibaba's various online marketplaces.[13] Its spider-like corporate campus, with its gardens and open-plan workspaces, spreads over more than 150,000 square metres in Ma's hometown of Hangzhou. It houses tens of thousands of employees, and the other businesses Ma has since founded. There is Taobao (meaning 'hunting for treasures'), launched in 2003. It is like eBay, enabling people to sell virtually anything directly to each other, but it is more like a local flea market where you might land on a bargain or find something bizarre, such as live scorpions or soap made from the seller's own breast milk. There are hundreds of 'Taobao' villages scattered across China where a large proportion of their economy is based on people selling local goods on the platform.

Then came Tmall in 2008, the online equivalent of a humungous glossy shopping mall, selling well-known global brands, from Disney to Burberry, directly to Chinese consumers. The Alibaba Group has 454 million annual active buyers, and one out of every three individuals in China has made a purchase from the marketplaces.[14]

On the day of Alibaba's Initial Public Offering (IPO), in September 2014, the dynamic Ma was ebullient about his company's historic milestone. In interviews he gave there was one word that stood out: 'Trust. Trust us, trust the market and trust the young people,' Ma said. 'Trust the new technology. The world is getting more transparent. Everything you worry about, I've been worrying about in the past fifteen years.' Ma continued without a pause. 'Trust has to be earned, of course. Because when you trust, everything is simple. If you don't trust, things get complicated.'[15] He made no fewer than eight mentions of the word 'trust' within the space of a minute.

Right from the start, Ma knew the importance of building trust, especially in a culture like China's – it's why he splashes the word about so generously. And we'll come back to how he set about doing just that. But first, it's worth unpacking the concept of trust itself. Like the rest of us, Ma would probably have trouble defining it – what exactly do we mean when we talk about 'trust'? And what doors does it open?

Trust is not a nicety, a kind of optional extra in life. We all depend on it in so many of our daily activities. How could we eat, drive, work, shop, get on a plane, go to the doctor, tell secrets, unless we trusted other people? 'Trust,' as political scientist Eric Uslaner says, 'is the chicken soup of social life.'[16]

For instance, when I order takeaway sushi, I have to trust that the restaurant will use fresh ingredients, the kitchen will be clean, they will not abuse my credit card details and the deliveryman will not run off with my dinner. Trust enables small and large acts of cooperation that all add up to increased economic efficiency. 'Virtually every commercial transaction has within itself an element of trust, certainly any transaction conducted over a period of time,' Kenneth Arrow, the American Nobel Prize-winning economist, once observed. 'It can be plausibly argued that much of the economic backwardness in the world can be explained by the lack of mutual confidence.'[17]

Trust enables us to feel confident enough to take risks and to open ourselves up to being vulnerable. It means we can commit to people before we know the precise outcome or how the other person will behave. And that applies to something as minor as ordering sushi or as life-changing as marrying someone. If, before we bought or did anything, we thought we were going to be cheated or ripped off, then very little would get done.

Social scientists, psychologists, economists and others view trust as an almost magical economic elixir, the glue that keeps society together and the economy ticking over. That much is agreed. The definition of trust, however, has been widely debated for years. In fact, there are more academic papers on its definition than on any other sociological concept.

It's odd that we describe trust as being *built* or *destroyed* when it's not a structure or a physical thing (except maybe in the form of a handshake or a paper contract). Like 'happiness' or 'love', it's one of those words that we tend to think of as a universal idea. And, the same as love, trust has many faces. It's not like some kind of predictable engine that comes with a manual and only functions in one precise way. Trust varies from situation to situation, relationship to relationship. Put simply, trust is highly contextual.

'So what does the word trust mean to you?'

It's a question I've asked hundreds of people over the past five years – entrepreneurs, politicians, leaders of large companies, scientists, economists, bankers, designers, academics, students and even five-year-olds. Their answers are fascinating and remarkably different. The question usually elicits a pause and then a kind of *mmm, I'm thinking* noise. 'It's hard to define, right?' they'll say. Yes, it is. Trust means different things to different people. The most honest answer I have ever been given was, somewhat ironically, from an insurance broker. 'Giving my wife my phone without clearing the history! That is trust,' he said.

For many people, it is about confidently relying on another person. For example, 'Trust is being able to rely on my husband/doctor/friends.' In this instance, trust is being referred to as an attribute that rests *in* someone specific, generally someone we are familiar with. The more we interact with a person over time, the more confident we become about how they will behave, that they are trustworthy. That kind of trust is known as *personalized* trust.[18]

Generalized trust is the trust we attach to an identifiable but unidentified group or thing. As one of the MBA students in my course at the University of Oxford's Saïd Business School put it, 'Trust is like a contract that guarantees an outcome.' For example, I trust the postal service to deliver my mail. It is also common to mix the two different types of trust. For instance, I might have high personal trust in my bank manager but fragile trust in my bank as a financial institution.

One of my favourite definitions of trust came from one of my son's friends. He was five at the time and had come round to our house to play. My son, Jack, told him over tea that I was writing a book. They were disappointed it was not about *Star Wars* or *Harry Potter* but still asked fascinating questions. I asked them what trust meant to them. 'Trust is when the ice-cream man says he will give you ice cream, he gives you the ice cream because he wants to, and I don't worry about him not giving me the ice cream,' replied my son's friend in one breath. Wow, out of the mouth of babes. It is in fact very close to the definition given by the pre-eminent German sociologist Niklas Luhmann, who wrote, 'Trust is confidence in one's expectations.'[19]

I am on the board of the National Roads and Motorists' Association (known as the NRMA), one of the most trusted brands in Australia. It's like the AAA in the United States or the RAC in the United Kingdom, basically the people who will come and fix your car if it breaks down, wherever you are. Recently, a woman phoned into the NRMA call centre. She sounded very distressed. She was breathing heavily and was clearly crying. Turns out, she was driving down the motorway when she realized she had passed the exact spot where her son had crashed and tragically died a few years ago. She pulled over on to the side of the road and started to have a panic attack. The first number she rang was the NRMA. A roadside assist personnel got to her within minutes. He sat with her for more than two hours. They listened to the radio together. They talked about her late son. He didn't leave until she felt ready to drive again. I was so touched by this story but also curious to find out, why did she call the NRMA? I mean, nothing was wrong with her car. Why not the police, an ambulance, her husband or a colleague? 'I knew you would come,' was her answer. That is trust.

Of the hundreds of definitions of trust I have studied, most can be reduced to one simple idea: trust is an evaluation of outcomes, of how likely it is that things will go right. Or put another way, trust is fostered when the likelihood of an undesirable outcome is low. The five-year-old was right, sort of. Five-year-olds tend to be more naturally trusting than grown-ups, having had a lot less experience in being let down or in worrying about distant outcomes. For adults, trust becomes more complicated and it works in the heart as well as the head. As Morton Deutsch beautifully put it, trust is 'confidence that [one] will find what is desired [from another] rather than what is feared'. Trust is a mixture of our highest hopes and our deepest worries.[20]

If you look up images for the word 'trust', all sorts of graphics will come up, often with some kind of danger lurking in the picture. People swinging between trapezes, for instance. Two hands, reaching out to meet but not quite touching. A person in mid-fall relying on another person – typically with their arms stretched out – to catch them. A sleeping lion with a mouse prancing a few centimetres from his nose. The common element in all of these areas is a gap, a grey area where something unknown happens. The images are conveying the powerful elements of trust: vulnerability and expectation.

Imagine a gap exists between you and something unknown. A stranger you need to rely on, a restaurant you have never been to before or your first run in a self-driving car. The gap between the known and unknown is what we refer to as risk. Indeed, risk can be defined as the management of uncertainty that matters. There are some uncertainties that are simply irrelevant. For example, if I am a farmer in England the possibility of heavy rain is an uncertainty that matters to my livelihood. But if I am managing a clothing factory in, say, China, the uncertainty about the weather in England is irrelevant. When there are no unknowns, when we can guarantee an outcome, there is no risk. For example, we know for certain the sun will rise in the morning.

Trust and risk are like brother and sister. Trust is the remarkable force that pulls you over that gap between certainty and uncertainty; as the Nike tagline says, 'Just do it'. It is literally the bridge between the known and the unknown. And that's why my definition of it is simple:

Trust is a confident relationship with the unknown.

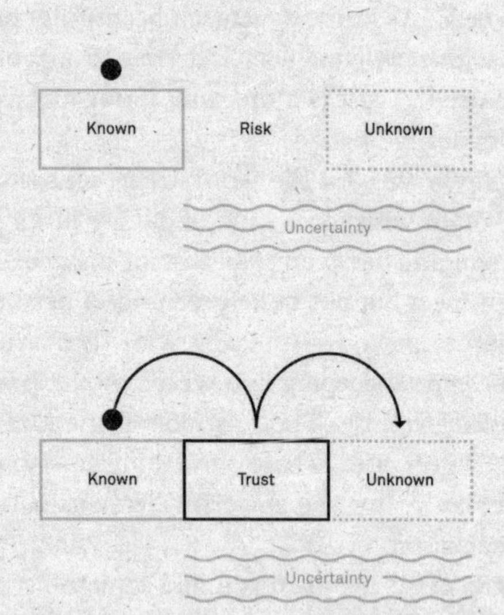

When you view trust through this lens it starts to explain how it enables us to cope with vulnerability, place our faith in strangers or just keep moving forward. It shows why just enough trust is a critical ingredient of innovation and entrepreneurial success like that of Jack Ma. Companies such as Apple, Amazon and Netflix constantly challenge assumptions, take smart risks and allow their employees to dive into uncharted waters to discover new ideas. But they also know how to get customers to trust new offerings, so that the initial risk of trying something new becomes quickly irrelevant.

Jack Ma realized the internet presented an opportunity to unleash the entrepreneurial spirit that existed in China but had been suppressed by years of Communism. What he spotted early on was how technology could enable trust – make unknown sellers seem familiar to people. But how to build a new kind of trust between strangers in a country based on *guanxi*?

Aside from those ancient traditions at work, less than 1 per cent of the country's population was online when Alibaba first launched. And of that 1 per cent, fewer still would even consider buying something off a website. People were unfamiliar with the concept of the internet, let alone e-commerce. Indeed, there was no history of e-commerce, no online payments system nor even a way quickly and safely to send goods. So how did Alibaba crack the trust code?

When we are trading goods over the internet, typically neither party knows each other. We are wary of scams and products not being as promised. If I buy, say, a Fitbit from an eBay seller, is it really brand new or is it refurbished, fake or even stolen? A lot can go wrong. There is always the chance of an unwanted outcome or a risk. Ma realized that to create trust between buyers and sellers online, he needed to use technology to reduce uncertainties or lower the risk enough to allow transactions to go ahead.

He also recognized that the bigger the trust problem, the greater the business opportunity. Ma was a bit like Steve Jobs; he knew there were enormous advantages to developing the solution to an obstacle that stood in the way of a market, rather than waiting for others to solve the problem. Take payments. How do I know you will pay? How do I know you will send what I paid for? It's a classic chicken-and-egg trust problem.

'For three years, Alibaba was just an e-marketplace for information. What do you have? What do I have? We talk for a long time, but don't do any business, because there is no payment,' says Ma. 'I talked to the banks. No banks wanted to do it. Banks said, "Oh no, this thing would never work," so I didn't know what to do.' Ma knew all too well that if he launched a payment system without the required licence, he would be breaking China's strict financial laws. He could have gone to jail. But he decided to do it anyway. Why? 'It is so important for China and the world to be able to trust the system,' says Ma.[21]

In 2004, the company introduced an online system called Alipay (支付宝, meaning 'payment treasure'). Instead of a direct payment like PayPal, Alipay takes money upfront from buyers and puts it in an escrow account.* The seller then ships the product and the funds are only released after the buyer has inspected the goods and confirmed that they are satisfied. It's a simple example of reducing uncertainties, in this instance over settlement. 'So many of the people I talked to at that time, for Alipay, said, "This is the stupidest idea you've ever had,"' recalls Ma. This was really stepping on the toes of China's highly regulated banking sector. But Ma didn't mind if people thought the idea was risky or even stupid, 'As long as people use it.' And use it they did. Today, more than 450 million people use Alipay to pay for goods.[22] It is estimated to be worth $74.5 billion as a standalone business.[23] In 2017, users sent $1.7 trillion in total payments through Alipay.[24] Ma always had big dreams.

Payments aside, how did people know they could trust the unknown small businesses and individual traders Ma wanted to put online? Take thirty-eight-year-old Wang Zhiqiang, one of the sellers who rang the bell on the day of the IPO. Zhiqiang was once a migrant rural worker struggling to make a living. He had tried his luck as a vegetable street vendor, a construction worker and a takeout deliveryman. Despite being poorly educated, Zhiqiang had always been interested in computers and the internet. He moved to Beijing's Zhongguancun area, known as China's Silicon

* An escrow account is generally a holding fund where money is held by a third party on behalf of two other parties making a transaction.

Valley, where he took on many manual labour jobs for more than six years. By 2006, he had saved enough money to buy a computer, which he took back to his hometown, a small rural farming village in north China's Shanxi province. After overcoming the challenge of getting the internet installed in his home, he opened an online shop selling just a few local products such as rice and soybeans. His friends and family wondered what on earth he was trying to do. In 2008, while the Olympic Games were in full swing in Beijing, Zhiqiang got the break he needed. He opened 'Farmville', not the popular game but an online shop selling all kinds of fresh produce grown by the local villagers. Sales quickly soared to more than 200 products per day. Today, his monthly net profits are more than 80,000 yuan (approximately $13,000) and he has become well known online as 'Wang Xiaobang' – in this context, 'bang' means a warm-hearted person who would like to help others.[25] But how did so many people come to trust the legitimacy of this unknown vendor from a remote area?

The answer is, at least in part, down to a service Alibaba launched in 2001, called TrustPass. For a seller to get TrustPass certification, they had to go through a third-party identity and bank account verification process. Alibaba also helped sellers create their own official-looking brands and virtual storefronts. Zhiqiang, for instance, would take colourful pictures of rural farm life and close-ups of the fresh products he sold as they were farmed or harvested. He wanted his store to feel 'local' and connected to the suppliers. TrustPass marked a breakthrough for Alibaba, not just in terms of trust but also money.

On average, certified TrustPass sellers were receiving up to six times more genuine enquiries than non-registered sellers. This gave Alibaba the perfect excuse to start charging small businesses (most services had been free up until this point). 'It made customers who paid for their status appear more *trustworthy*,' says Porter Erisman, a close friend of Ma and long-time employee. 'It made those members still clinging to their free accounts seem *less* trustworthy. After all, if they had such a good business, why weren't they willing to pay up a little to prove it?'[26]

Alibaba's liquid gold was not online shopping but trust. And that is why, when Ma discovered that it had been badly broken, he was livid. In February 2011, it became known that around a hundred members of Alibaba's 5,000-strong sales team had been taking financial kickbacks in exchange for allowing fraudulent sellers to skip the verification process and set up accounts. The deceitful behaviour had been going on for more than two years. The impact? Two thousand three hundred and twenty-six high-volume sellers of low-quality or even fake products had been verified as 'Gold Suppliers'.[27]

Ma knew he had a grave trust issue on his hands. He had to act quickly. He had to send a loud and clear message to protect his company's reputation. So the salespeople who had knowingly set up the accounts and those employees who had looked the other way were all fired. It was a well-managed *trust breach*, and Ma's actors David Wei, the then CEO, and COO Elvis Lee also resigned, even though they were not implicated in the fraud. Both were falling on their swords to accept responsibility. 'One of our most important values is integrity. That means integrity of our employees and integrity of our online marketplaces as trusted and safe places for our small business customers,' said Ma. 'We must send a strong message that it is unacceptable to compromise our culture and values.'[28] His actions were clearly validated. In 2016, the Alibaba Group overtook Walmart as the world's largest retailer.

Ma has proved in spades to the Chinese (and to the rest of the world) that trade does not require a prior or close personal tie in order to work. And strangers won't, as a matter of course, betray you.

The story of Alibaba is a telling illustration of how technology is enabling millions of people across the world to take a *trust leap*. A trust leap occurs when we take a risk and do something new or in a fundamentally different way.

Trust leaps create new possibilities; they break down barriers and help us form new relationships; they enable us to mash up ideas and memes in unexpected ways; and, as in the example of Alibaba,

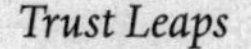

TRUST LEAPS

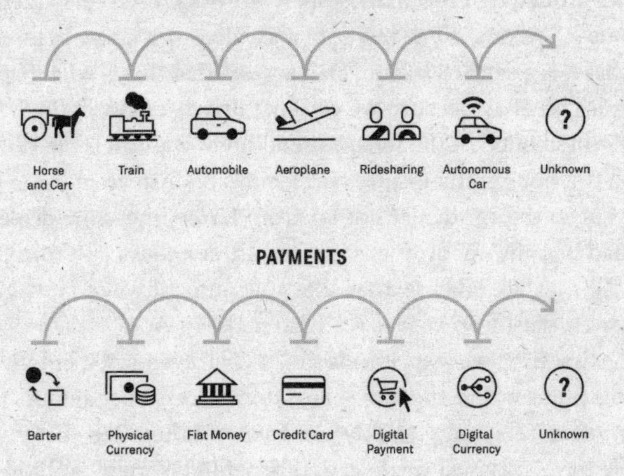

to open up new markets, new networks and new alliances that would once have been unthinkable. Trust leaps carry us over the chasm of fear, that gap between us and the unknown.

Imagine the first time people switched from bartering real goods to using paper money. Bartering is intuitive – I give you a chicken in exchange for a metal pot. Money meant people had to trust that these flimsy pieces of printed paper had real value, and would retain that value. They had to trust that the institutions issuing the money, usually governments and banks, would determine the right value. That is a trust leap.

Do you remember the first time you put your credit card details into an internet site? That is a trust leap.

I remember a heated conversation I had with my dad when I was eighteen. I had seen a nice-looking second-hand navy blue Peugeot for sale on eBay. It was within the price range my parents had set me to buy my first car. From the photos, it looked in good condition. My dad, then a chartered accountant, asked me if I knew what 'a market for lemons' was. I didn't back then. Over lunch, a mini lecture followed on George Akerlof's economic theory about problems of uncertainty over quality of goods.[29] To put it simply, Akerlof argued that in a market there are good used cars and defective ones ('lemons'). The buyer of a car does not know beforehand whether it is a car or a lemon. Dad argued that this is what happens in spades on eBay because we couldn't drive or inspect the car. He also pointed out that the seller's pseudonym was 'Invisible Wizard', which did not exactly inspire confidence. So instead of using eBay, we went to the car dealer not far from home, the same dealer my dad had bought my brother's car and three other cars from in the past. To my dad, eBay seemed like an irrational way to buy goods. It wasn't a trust leap he was ready to make in 1999.

The first time we leap, it feels a bit weird, even risky, but we soon get to a point where the idea seems normal. Our behaviours transform, often relatively quickly. And when others see that enough people have survived the leap and benefited from it, millions will follow. My father is now somewhat of an eBay addict. He would probably consider buying a car on eBay these days (then again, maybe not).

Trust is the conduit through which new ideas travel. Trust drives change.

Throughout history, humans have demonstrated a remarkable propensity to change the way we do things – bank, trade, travel, consume, learn and date. To understand just how good we are at taking trust leaps we need to go way back in time, way before the internet or even the printing press were invented.

One day in the year 1005, a handwritten letter crossed the Mediterranean. It was filled with the troubled words of Sumhun ben Da'ud, an eminent merchant who lived in Sicily. He was furious with his business partner, Joseph ben 'Awkal, who had ignored Sumhun's repeated requests to pay Egyptian creditors the hundreds of dinars they were owed. As the total monthly expenses of a middle-class family living in Egypt were no more than three dinars, these were very significant sums. Word was spreading to other traders across the region that neither of the business partners could be trusted. 'My reputation is being ruined,' lamented Sumhun.

Sumhun and Joseph were part of a close-knit group called the 'Maghribi traders', Jews who left Baghdad during the political turmoil of the tenth century and settled in the Maghreb, the coastal region of north Africa. Around 150 years ago, more than 1,000 of their personal letters were discovered in almost perfect condition in the storeroom of an ancient synagogue in Fustat in Egypt.[30] The letters provide an intriguing window into the lives of the traders and the role they played in transforming long-distance trade.

Say a merchant based in Old Cairo wanted to sell his textiles and spices in Palermo in Sicily. He could travel long distances by boat – but sea voyages were treacherous and time-consuming. Or, instead of travelling himself, he could use overseas agents, and the agents could handle everything from unloading the ships to selling the goods in local markets, to settling the odd bribe on the merchant's behalf.

Today this sounds like a relatively straightforward idea, but at the time it required a massive trust leap. Possibilities for deception and corruption were high. An agent could lie about prices, skim money or simply steal the goods altogether. And when something bad happened, it could take months for the merchants to find out. Formal trade regulations and legal contracts, as we know them, did not exist. The Maghribi merchants faced a problem: they had no way of knowing what the hired agents were up to on the other side of the Mediterranean.[31]

When one party has less information than the other, economists call it *information asymmetry*. Economist Kenneth Arrow first described the concept in 1963, in the context of healthcare. Doctors generally know more about the value and effectiveness of a given medical treatment than patients do. They are in a powerful 'expert' position and patients will tend to follow their recommendations. Arrow noted how sometimes the doctor might manipulate the asymmetry to his or her advantage, for instance, recommending costly drugs or an operation that is not necessary.

Information asymmetry is all around us because it is rare for two people to have perfect and equal information in any kind of exchange. The life insurance broker who knows what clause 221 of a complicated policy really means; the Alibaba seller Rakjuk Kft who knows if the top-quality Angora goats he sells really are 'well-bred' champions, 'free from parasite' and 'fully red-blooded', as advertised; the second-hand car salesman who knows the real history of a cute little Fiat 500; the Airbnb host in Cape Town who knows if their place really has two bedrooms (or if one is a sort of pseudo-loft in the kitchen only accessible by a precariously propped-up wooden ladder) and the sweeping ocean views shown (or has a postage-stamp view of something blue if you hang off the right-hand corner of the balcony and use binoculars); and the overseas agents who know if they are selling the merchant's frankincense or olive oil for fair or dodgy prices. Information asymmetry creates future unknowns and the very need for trust.

So how did the Maghribi traders get the agents in far-off lands not to lie, cheat or steal in the absence of direct supervision? The system they came up with was so ingenious that it opened the modern era of long-distance trade between strangers.

The Maghribi traders had the same religion, common family ties and, most importantly, shared the same motive of ensuring their agents did the right thing. They were living in the era of *local trust* and had high levels of *social capital*. It's an idea that has fascinated sociologists for centuries, notably Pierre Bourdieu, Robert Putnam and James Samuel Coleman. 'Whereas physical capital refers to physical objects and human capital refers to the properties of individuals, social capital refers to connections among individuals – social networks and the *norms of reciprocity* and trustworthiness that arise from them,' Putnam explains in his influential book, *Bowling Alone: The Collapse and Revival of American Community*, first published in 2000.[32] Putnam argues that close-knit suburbs have given way to 'exurbs' and 'edge cities' – vast anonymous places where people spend more time commuting to work, at the office and watching TV alone, and less time socializing with friends, neighbours, community groups and even family.

Social capital – shared values, bonds and support – can be found in a whole array of networks and communities. For instance, a group of neighbours who keep an eye on one another's homes to make the street safer. A school that holds a garage sale to raise money for a local homeless shelter. A stranger who drops your wallet into the police station after finding it left on a bus. A former colleague who helps you land a new job. 'Out of such shared values comes trust, and trust has a large and measurable economic value,' writes Francis Fukuyama in his 1995 classic book *Trust*.[33]

The merchants formed a coalition with a system of collective sanctions. The tight-knit group would frequently write and talk to each other, openly sharing information about good agents and those up to no good. However, they couldn't just rely on shaming the cheaters. Merely identifying them was not enough to discourage bad behaviour. Rewards for acting honestly and responsibly were needed as an incentive against the short-term gains cheating offered.

The Maghribi traders designed a simple reputation-based system where the most trustworthy agents were rewarded with the most business. If one agent ripped off a merchant, the entire trading network subsequently shunned him. All the merchants were required to take a collective vow never knowingly to employ crooks. The agents knew their success in the long run depended on repeat business. The traders were able to set aside their fears of getting scammed because both parties knew honesty would pay. (That was why Sumhun ben Da'ud was so upset with his business partner.)

What's more, it turned out that it didn't matter if the merchants weren't aware of every wrongdoing; the threat that they could find out was enough to make the agents behave honestly. If people believe they are being observed and judged, it makes them behave better, even if they are not actually being watched all the time.

Just like Jack Ma would do a thousand years later, the traders created a system to reduce the unknown enough for people to take a risk and do things differently. Through simple accountability mechanisms, they were able to expand their local network into international trade across geographic, language and cultural boundaries.

Trust leaps expand what is possible, what we can invent and who can be an inventor. Trust leaps extend the reach of our collaboration and creations, opening up new horizons of opportunity. That is why trust matters *so* much and why establishing confidence in the unknown has been a central part of innovation and economic development over the course of history. Just ask Jack Ma, the boy who was told he was no good for anything, who took a trust leap on himself and persuaded a nervous nation to leap with him.

2

Losing Faith

Chances are you have never heard of an American woman called Jean Heller. She was a young reporter just a couple of years out of graduate school when she uncovered a dark secret the United States government had been hiding for more than four decades.

It was this. Between the years of 1932 and 1972, 600 African Americans living in Tuskegee, one of the poorest counties in rural Alabama and with the highest syphilis rates in the nation at the time, had been used as human guinea pigs by the United States Public Health Service. It was one of the most unethical medical research experiments the country had ever seen and would become known – rather blandly, in light of what it represented – as the 'Tuskegee Study'.

In the experiments, the county's black farmers, many illiterate, were subjected to painful spinal taps, daily blood draws and, when they eventually died, autopsies. The men were offered free hot lunches, transportation to and from hospitals, free medicine and free burial services as incentives to enter the study programme.[1] The farmers were never told what disease they were suffering from or of its seriousness. Of the 600, 399 of them were syphilitic, while 201 did not have the disease but were used as controls. Doctors merely informed all participants they were being treated for 'bad blood'.[2] These trusting men, disadvantaged and easily manipulated, became pawns in what James H. Jones, author of *Bad Blood*, identified as 'the longest nontherapeutic experiment on human beings in medical history'.[3]

Shockingly, the farmers were left to suffer the ghastly ravages of syphilis, which can include blindness, deafness, dementia, heart disease, paralysis and eventually death. And not knowing they had the disease, the men unwittingly passed it on to their wives and children. 'The Tuskegee Study began ten years before penicillin was found to be a cure for syphilis and fifteen years before the drug became widely available,' wrote Heller in her story published on 26 July 1972, in the *New York Times*. 'Yet, even after penicillin became common, and while its use probably could have helped or saved a number of the experiment subjects, the drug was denied to them.'[4]

Why? Because the very purpose of the study was *not* to cure the participants of syphilis. The goal was to observe the long-term effects of the disease and to determine through autopsies if untreated syphilis affected black bodies the same way it affected white ones. 'As I see it,' one of the doctors explained, 'we have no further interest in these patients until they die.'[5]

Heller's story became front-page news and generated widespread public outrage. Members of the US Congress reacted with shock, denying they had known what was going on. The egregious research was stopped immediately, but the fact that it had continued over four decades was the result of a lot of people turning a blind eye for a very long time.

Responding to the outcry, congressional investigations led to the establishment of the Office for Human Research Protections, to monitor ethical standards. Federal laws were also created, requiring Institutional Review Boards to oversee clinical research and ensure adequate protection of all study participants.

Twenty years later, the government, right from the top, was still apologizing for the 'moral and ethical nightmare' of Tuskegee. 'It was a time when our nation failed to live up to its ideals, when our nation broke the trust with our people that is the very foundation of our democracy,' President Bill Clinton said in a press conference in 1997, standing alongside eight elderly survivors of the study. 'To our African-American citizens, I am sorry that your federal government orchestrated a study so clearly racist. That can never be allowed to happen again . . . An apology is the first step, and we take it with a commitment to rebuild that broken trust.'[6]

This shameful chapter in medical research history shook the foundations of trust between Americans, especially black patients, and the medical system for a long time afterwards. It's only recently, however, that anybody has tried to quantify the scale of the trust fallout.

African-American men today have the worst health outcomes of all major racial, ethnic and demographic groups in the United States. For black men at age forty-five, the life expectancy is three years less than for their white counterparts.[7] It is caused by multiple factors, including disparities in income, diet and healthcare access. But could the differences also be linked to the general African-American distrust of healthcare providers that grew out of the Tuskegee Study? And if so, to what degree?

These are questions that intrigued two researchers, Marcella Alsan at the Stanford Medical School and Marianne Wanamaker, an economist at the University of Tennessee. After crunching data from various national population and health surveys, in 2016 the pair confirmed what other researchers had hypothesized: that high levels of mistrust existed among African-American men *decades* after the Tuskegee Study was exposed.[8] The difference was that Alsan and Wanamaker put a precise number on its life and death effect.

When someone doesn't trust their doctor, typically they stop going for check-ups or for care when they need it. The researchers deduced that, as a general principle, this has led to a decrease in life expectancy of 1.4 years among black men over forty-five.[9] Perhaps the most remarkable finding, however, was that more than a third of the life-expectancy gap – between older black men and their white equivalents – could be attributed to fallout from the disclosure of the Tuskegee experiment. It's a staggering revelation: life expectancy has dropped for millions of men who didn't live in Alabama and had nothing to do with the Tuskegee Study because of broken trust – or to put it another way, compounding mistrust.

Dr Joseph Ravenell gave a wonderful TED talk in 2016 on this very problem that continues today. He pointed to the fact that high blood pressure is one of the leading causes of death among African males over fifty, a medical problem that could be prevented with timely diagnosis and appropriate treatment. So why is it so deadly for black men? 'Because too often high blood pressure is either untreated or under-treated in black men, in part because of our lower engagement with the primary healthcare system,' says Ravenell. 'Some of our earliest research on black men's health revealed that for many, the doctor's office is associated with fear, mistrust, disrespect and unnecessary unpleasantness.'[10] So people skip going to the doctor, especially if they feel fine. Indeed, Ravenell has found in his research that many black men trust their barbers far more than they trust their doctors.

It doesn't matter that regulations and ethical standards have since been put in place; the echoes of Tuskegee still linger and affect some people's decisions.[11] The findings also tell a bigger story: they show how one trust-busting incident can create a generational scar against an institution or system that takes decades to heal. And sometimes the damage is too deep to repair.

It would be comforting to think that trust-shattering incidents like the Tuskegee Study were all in the past, the lessons learned and acted upon. The truth is that institutional trust is in greater jeopardy now than ever before. For a stunning illustration of why ordinary citizens feel more betrayed than ever by elites and those in power, and why those elites themselves are heading into a twilight zone, we need only look to the Panama Papers.

It began with a message – anonymous, of course. 'HELLO. This is John Doe. Interested in data?' The recipient was Bastian Obermayer, a thirty-eight-year-old investigative reporter for the German newspaper *Süddeutsche Zeitung*. It was sometime late in 2014 when the mysterious message appeared on the reporter's laptop. Obermayer was sitting in his flat in a quiet neighbourhood in Munich at the time. 'We're very interested,' he quickly typed back.

The source replied in encrypted chat that he or she was going to leak a mountain of highly confidential data, 'more than you have ever seen'. The source made it clear upfront that he didn't want money, only justice. 'I can't explain my rationale without making my identity clear . . . but I want to make these crimes public,' Doe replied.

Over the course of 2015, more than 11.5 million of the documents that became known as the Panama Papers were transferred. They represented forty years' worth of digitized records ripped from the servers of the Panamanian law firm Mossack Fonseca. The amount of data was staggering; we're talking 2.6 terabytes, and it included 4.8 million emails, 2.1 million PDFs, 3 million database files, 1 million images, as well as other confidential contracts, letters, bank records and property titles.[12] The documents would turn out to be the greatest single data drop in journalistic history, 2,000 times larger than the 2010 Cablegate coup, when WikiLeaks released more than 250,000 classified diplomatic cables that had been sent to the United States from 270 consulates and embassies around the world. These cables contained allegations of corruption and revealed numerous unguarded comments such as US embassy staff referring to Vladimir Putin as an 'alpha-dog', Hamid Karzai as being 'driven by paranoia' and a comparison of the then Iranian President, Mahmoud Ahmadinejad, to Adolf Hitler.

The Panama Papers revealed that between 1977 and 2015, the firm Mossack Fonseca had created more than 200,000 offshore shell companies in tax havens around the world, for world leaders and their families including President Putin and Kojo Annan, son of the former UN secretary-general Kofi Annan, celebrities such as footballer Lionel Messi, and many other important global elites. But beyond the revelations it laid bare, the leak itself represents a remarkable trust story.

Imagine you have been handed an immense cache of secret documents. What would you do?

Journalists have, by nature, an obsession with scoops, with being first. But after working non-stop for more than two months, Obermayer realized he was in way over his head with the small team at his German newspaper. The volume of data was daunting. He needed help, lots of it. So he took the unusual step of sharing the files with hundreds of other journalists from around the world.

Obermayer first reached out to Gerard Ryle, someone he deeply respected and had worked with before on the Luxembourg and HSBC leaks. Ryle is the director of the International Consortium of Investigative Journalists (ICIJ), a non-profit based in Washington, DC. It was set up in 1997 to create a network of journalists around the world to expose global scandals. It was, for example, the ICIJ who first revealed the war contracts in Iraq and Afghanistan that were unfairly awarded to companies that had donated large sums to the presidential campaigns of George W. Bush.

Ryle is an Irish-born reporter. He is a calm and softly spoken man. He is the kind of person you can imagine not talking much at a dinner party but saying the one thing everyone remembers. A veteran newshound, he is the embodiment of an investigative journalist, ready to shake every tree and go down every burrow to find the truth, to make sure he gets a story right. He became Director of the ICIJ in 2011.

In June 2015, Ryle and Obermayer organized a secret meeting on the thirteenth floor of the National Press Club in Washington. Staff from rival news outlets like the *Guardian*, BBC, *SonntagsZeitung* and *Le Monde* gathered around a long conference table.[13] During the meeting, the journalists decided the project needed a code name. Prometheus was agreed upon, after the Titan from Greek mythology who stole the secret of fire from the gods. Before they could get access to the documents, the journalists had to sign a basic non-disclosure agreement. It required them to operate under only two rules. First, to share everything they found, and second, to publish on the same day, at the same time, and Ryle had the power to set the publication date.

All the leaked documents were stored in a searchable database program named Blacklight. If you wanted information on a specific name, for instance 'Ian Cameron', the database would run a search and in a matter of minutes produce a CSV file with all the matching documents. Some of the clients, however, were harder to identify – Mossack Fonseca addressed some of the more secretive ones by code names like Winnie the Pooh or Harry Potter.

Working collaboratively, local journalists – from Iceland to Nigeria to Russia – could piece stories together in ways that would have been almost impossible to do from a newsroom in one country. 'While one journalist is looking at Indian data, it might lead them to Brazil, or France, and then suddenly you've got this trust-building exercise where the Indian journalist shares information with the French journalist,' explains Ryle.[14]

A critical member of the ICIJ leadership team was thirty-three-year-old Spanish-born Mar Cabra who heads data and research. Her colleagues describe her as a 'data genius'. Mar distinctly remembers the day a colleague discovered Messi among the client list. 'A message appeared in the virtual newsroom saying, "Oh my god, I found the football player Lionel Messi!" The other journalists were like, "Oh my god, Messi."' Once one person had found Messi in the documents, everybody had found Messi.

After nine months of painstakingly combing through tens of thousands of documents, the journalists were exhausted and stressed, and champing at the bit. Many times, Ryle had to calm jittery journalists whose natural instinct was to start reporting on the injustices they had discovered. 'It took a lot to keep everyone holding the line,' he says. The journalists were spread out across time zones, working in different languages and based in countries with different social and political environments. Ryle was not technically their boss and he couldn't force them to do anything they didn't want to do. At the end of the day, everyone was legally free to report as they saw fit. Yet, remarkably, the story remained a secret for nearly a year.

Ryle knows a thing or two about what makes journalists tick. 'A journalist's biggest weakness, whether they like it or not, is their ego. But on this project, we had an awful lot of selflessness going on,' he says. 'I tried to give them a sense of belonging to something greater than themselves.' The prospect of a bigger and better story if they all worked together was used as a carrot, but Ryle knew he also needed a stick. 'I made it clear that if they broke the trust once, the ICIJ will never work with them again.'

Ten minutes before the agreed embargo was due to expire, an unexpected message appeared on Twitter: 'Biggest leak in the history of data journalism just went live, and it's about corruption.' It was tweeted by Edward Snowden. Juliette Garside was sitting in the offices of the *Guardian* with her bosses and the rest of the Panama Papers team when it came. 'We had our fingers hovering over the button to go live with our story but we were respectfully waiting for the exact embargoed time,' she says. 'Then the Snowden tweet appeared and all hell broke loose. Everyone in the room was shouting "Go, go, go!"'

On Sunday, 3 April 2016, at 8 p.m. German time, dozens of news organizations around the world started publishing front-cover stories about the Panama Papers. Just hours after publication, thousands of people took to the streets of the Icelandic capital of Reykjavik to express their outrage and to call for their prime minister to quit. The papers revealed that Sigmundur Davíð Gunnlaugsson had a secret offshore tax account called Wintris Inc., in the British Virgin Islands, that he used to hold investments with his wealthy wife. He reportedly avoided more than 500 million Icelandic krona (approximately $4 million) in taxes on the wealth. To make matters worse, he had set up the account in 2007, while Iceland was on the brink of a financial meltdown. Gunnlaugsson resigned shortly after the leak.[15]

Bastian Obermayer, the whistle-blower's original contact, was one of the first to discover the links to Vladimir Putin in the files. He repeatedly came across the name Sergei Roldugin, a Russian concert cellist. It turned out that the cellist had been a friend of Putin since they were teenagers and was godfather to the president's daughter, Mariya. It was reported that Roldugin owned a Moscow bank called Bank Rossiya that had allegedly handled billions of dollars in transactions with offshore companies.

In all, 29 of the billionaires featured in *Forbes Magazine*'s list of the world's 500 richest people, 12 serving or former world leaders and 140 politicians were named in the Panama Papers. The documents implicated the king of Saudi Arabia; Ian Cameron, the late father of former UK prime minister David Cameron; six members of the House of Lords; Nawaz Sharif, Pakistan's prime minister; Ayad Allawi, former vice president of Iraq; and Petro Poroshenko, president of Ukraine.

In itself, the appearance of a name in the files was no proof of wrongdoing; there are plenty of legitimate reasons to hold money in offshore companies and trusts. A foreigner, for example, buying a vacation home, may set up a local shell company to purchase the property. Indeed, Mossack Fonseca vigorously denied having broken any laws, as did many of its clients. Even so, the avalanche of information provided damning examples of how the rich and powerful can exploit offshore tax regimes and strip countries of tax revenues that most believe they deserve to pay.

The public outrage was loud and vehement. Here were the rich finding ways to grow even richer by avoiding taxes the average person must pay. Here were leaders essentially salting away money for which tax should morally be payable for the benefit of the very countries they were governing.

Was this all we could expect from those at the top level of society, those who are meant to serve the public good?

The feeling of disillusionment that followed the leak wasn't just about money; it was about fairness and equality. Why did the wealthy, powerful and elite get to play by different rules? The revelations left the social contract in tatters: it destroyed the tacit understanding that we all work hard, pay our taxes and are 'in this together'.

In revealing to the average person what went on behind the scenes, the Panama Papers confirmed one of the key reasons why institutional trust is eroding at an alarming rate; many people feel let down and left behind, watching in dismay as the elites and authorities in charge seem to thrive and act unethically, at the very least. To borrow the title of a book by Joseph Stiglitz, the Nobel Laureate economist, trust is facing a Great Divide.[16]

Institutions, as the eighteenth-century Italian political philosopher Giambattista Vico observed, are essentially social structures made up of a history of practices, values and laws that are accepted and used by many people.[17] We often think of institutions as something physical – grand university buildings, ancient stone churches, the Houses of Parliament – but they can also be an idea, constraint or a social norm. Marriage, for example, is an institution. So is the family unit or the British monarchy. Religion, property rights or other legal constitutions are institutions. 'The simple point is that institutions are to humans what hives are to bees. They are structures within which we organize ourselves as groups,' writes the historian Niall Ferguson in *The Great Degeneration*. 'You know when you are inside one, just as a bee knows when it is in the hive. Institutions have boundaries, often walls.'[18] In other words, concrete or conceptual, they are valued building blocks of rule and repetition on which societies are built. They shape our behaviour and how we interact with each other.

Trust in institutions bobs and dips with scandals, recessions, wars and changes in government. In the past, it was much easier to hide wrongdoings, such as Tuskegee, for years, even decades. Now, in a digital age, shoddy institutions and the leaders and elites at their helm are much more likely to be exposed and lose our trust quickly, sometimes for good. A deep loss of faith in banks, governments, the media, the church or other elite institutions is not a new phenomenon. Go back to any other civilized age and you'll find examples, such as the Crédit Mobilier Scandal of 1872 where a sham company was used as a front to bribe US congressmen and to funnel money to its construction projects. The scandal became a symbol of post-Civil War corruption. And then, a couple of years later, there was the Whisky Ring, exposed in 1875, involving hundreds of politicians, distillers and distributors who conspired to avoid payment of taxes by reporting lower alcohol sales. Unprecedented, however, is the extent and rate of the breakdown of trust we are now witnessing between citizens and institutions, between the everyman and the elites. Alarmingly, survey after survey of public sentiment across countries and age groups tells a similar, woeful tale.

In the 1970s, post-Watergate and the Vietnam War, when trust in government and the armed forces had slumped, Gallup began asking Americans how much confidence they had in their major institutions, such as banks, media, public schools, organized religion and Congress. Approximately seven in ten Americans believed they could trust key institutions to do the right thing most of the time.[19]

Over forty years on, the same Gallup survey continues. In 2017, it revealed that the confidence in fourteen institutions averaged at only 35 per cent. Confidence has fallen to historic lows across every single major institution, bar two – small business and the military.[20] When the survey first started, 75 per cent of Americans said they had 'a great deal or a fair amount of' trust and confidence in the federal government in Washington in handling international problems, and 70 per cent had confidence in the government for handling domestic problems. Those figures are now at 52 per cent and 54 respectively.[21] For Congress, the numbers are even worse, plummeting from 42 per cent in 1973 to 9 per cent now.[22] Even the Supreme Court, once a bastion of trust in society, has suffered a major decline – from 45 per cent in 1973 to 40 per cent today.[23] But it's not only governmental organizations where trust has eroded. Public faith has also taken a hit when it comes to banks (60 to 34 per cent);[24] big business (26 to 21 per cent);[25] the church (65 to 41 per cent);[26] and newspapers (39 to 27 per cent).[27]

In terms of age groups, millennials are the most doubting. According to a 2015 survey conducted by the Harvard University Institute of Politics, 86 per cent of them distrust financial institutions. Three in four millennials 'sometimes or never' trust the federal government to do the right thing and a staggering 88 per cent 'sometimes or never' trust the media.[28]

The plunge isn't limited to the United States. The story is similar across Western Europe and Britain. Respected pollsters Ipsos MORI have tracked people's trust in twenty-four different occupations in the United Kingdom, from politicians to hairdressers, for more than thirty years. Nurses come out on top as the most trusted, with a stellar rating of 94 per cent. But in this kind of survey, the category to watch is the one with the sharpest decline. It shows how easily once mighty institutions can crash and burn. According to Ipsos, the big loser this time round was the clergy. When the Ipsos poll started in 1983, 85 per cent of the people trusted the clergy to tell the truth. It was the most trusted profession. By November 2017, the clergy had fallen 20 percentage points to come in as only the tenth most-trusted profession overall.[29] Consider this: the average Briton now trusts the random stranger they meet on the bus or in a supermarket to tell the truth more than they trust a member of the clergy on the other side of a confessional.

So why is trust in so many elite institutions collapsing at the same time? There are three key, somewhat overlapping, reasons: inequality of accountability (certain people are being punished for wrongdoing while others get a leave pass); twilight of elites and authority (the digital age is flattening hierarchies and eroding faith in experts and the rich and powerful); and segregated echo chambers (living in our cultural ghettoes and being deaf to other voices).

For institutions to retain credibility and our confidence, there must be penalties – loss of power or position, fines – when they break the rules or the rules become meaningless. Take something clear-cut like traffic regulations. In Britain, the legal convention is to drive on the left. If over the course of, say, a week, we saw hundreds of cars veering down the wrong side of the road without any consequences, the power of the traffic rule would quickly dissolve. Similarly, rogues within institutions must be seen to pay or be punished. When they get off scot-free, our faith in the institution is shaken. And in recent years, we've seen many leaders get off scot-free. Consider banks. How did the crash of 2008 – the largest man-made economic catastrophe since the Depression – result in the jailing of only a single investment banker and minimal reform of Wall Street? What did that do to our trust?

Over the past couple of decades, the banking industry has had its skirts lifted to reveal some very grubby underwear. From Enron to Arthur Andersen, Freddie Mac to Fannie Mae, Lehman Brothers to Bear Stearns, AIG to Northern Rock, Nick Leeson to Bernie Madoff, the BHS pension funds fiasco to the Libor scandal, the list goes on, and it has taken a hard toll on trust. Perhaps the biggest blow, though, has come from the fact that only a handful of CEOs, the 'captains of finance' who played a role in creating the financial crisis, faced any form of punishment. The few who lost their jobs, most notably Ken Lewis of Bank of America and Dick Fuld, the former CEO of fallen Lehman Brothers, walked out the door with multimillion-dollar golden parachutes. The message is clear: if you are rich and powerful, you can break the rules, as long as it makes a lot of money. It's an acute case of moral hazard; when things went belly up, the bankers didn't face any real consequences.

Some bankers, such as Madoff and Leeson, proved inherently untrustworthy. But for the most part, banks are not filled with bad people; it's more a case that people working in banks are operating in a toxic culture with a perverse incentive structure that permits – and even breeds – unethical behaviour and misaligned interests. They acknowledge as much themselves.

Labaton Sucharow, a respected law firm based in the United States, conducted an independent survey to find out how financial insiders view other professionals within their industry. More than half of respondents believed that their competitors engaged in illegal or unethical behaviour. Nearly a quarter admitted they would engage in insider trading if they could get away with it and just under a third believed that financial services professionals might need to engage in illegal or unethical behaviour to be successful.[30] 'The succession of scandals means it is simply untenable now to argue that the problem is one of a few bad apples,' admits Mark Carney, the governor of the Bank of England. 'The issue is with the barrels in which they are stored.'[31]

Rising star Andy Haldane, a Yorkshireman, is currently chief economist at the Bank of England, working alongside Carney. He has become a key figure in the debate on financial regulation calling for 'reformation'. Indeed, in 2014, *TIME* named Haldane as one of the 100 Most Influential People in the World for being 'the central banker not afraid to be blunt'.[32] 'The significance of these findings is not the precise percentages, as striking as these are,' says Haldane in response to the Labaton findings. 'More fundamentally, it is because of what they reveal about finance's perception of itself, the mirror it holds to the social identity of finance . . . It is the sociology and psychology of banking and bankers that needs to change, as much as their finances.' The problem is, can you regulate culture?

Faith in the financial institution will not be restored unless the behaviour of banking changes and we see more in the way of serious punishment or penalties for those in the rotten barrel. The systemic breakdown of trust in financial institutions comes down to this application of different rules for different folks. 'Along with the other rising inequalities we've become so familiar with – in income, in wealth, in access to politicians – we confront now a fundamental inequality of accountability,' writes American political commentator and author Christopher Hayes in his fascinating book *Twilight of the Elites*. 'We cannot have a just society that applies the principles of accountability to the powerless and the principle of forgiveness to the powerful.'[33]

When it was revealed in 2015 that over 11 million Volkswagen vehicles were knowingly programmed with software, so-called 'defeat devices', which can dupe government emissions tests, CEO Martin Winterkorn resigned. His pay-out post 'dieselgate' was a staggering 15.9 million euro.[34] Likewise, no individuals or institutions have taken a serious hit following the 2016 Sir John Chilcot inquiry, which found that intelligence about weapons of mass destruction (WMDs) had been misrepresented in order to justify the invasion of Iraq following the attacks on the World Trade Center. I distinctly remember, in 2003, watching the news as the 'war on terror' began, bombs lighting up the Baghdad skyline. More than a decade of military interventions followed, unleashing astronomical human destruction and massive financial costs. When the long-awaited Chilcot findings, a 2.6 million-word report, were published in June 2016, the verdict was unequivocal: the legal basis to invade Iraq was 'far from satisfactory'. The report confirmed that the 'intelligence' about Saddam Hussein's alleged WMDs was 'flawed' and exaggerated.[35] The British prime minister's justifications may have destroyed the trust of the British voters but no politicians from the time, including Tony Blair and President Bush, were held truly accountable. It hasn't helped that both those leaders have been defiantly unapologetic.

As professors Alsan and Wanamaker proved with the Tuskegee Study, specific events can trigger rampant mistrust of entire systems. If politicians can take us into war under false pretences, we start to wonder how we can trust the wider process of decision-making in government. If Chuck Blazer, the ex-member of FIFA's advisory committee, accepted bribes, corruption must be pervasive in other sporting organizations. If we can't trust the behaviour of bankers, the financial system must be broken. If we can't trust journalists to report accurately on the financial crisis, the Iraq War or the presidential race, the press must be failing us as a credible source of facts. If we can't trust the Catholic clergy to report abuse, perhaps it means the leaders of the church are only loyal to the institution, not the people they are meant to serve.

It's as if the safety net we once relied on – our trust in the wider society and its sterling institutions – has been ripped away and we're in a spiralling trust freefall. It's not only that corrupt individuals are getting away with bad deeds; as the Panama Papers showed, the moral compass at the top of society is also spinning wildly.

It is easy to see why 'loss of trust' in established institutions has become both a mantra and a real crisis of our times. The problem is further amplified by the fact that many of us are trapped in echo chambers of information shared by 'like-minded' people, where we hear this message over and over.

On 29 June 2016, Facebook made an announcement about changes it was making to its personal news-feed algorithm. 'We are updating News Feed over the coming weeks so that the things posted by the friends you care about are higher up in your News Feed,' wrote Facebook's engineering director Lars Bäckström.[36] The statement sounded fairly innocuous but what it meant was significant; the feed would now promote content posted by friends over content by traditional media outlets.

Given that an estimated 41.4 per cent or more of traffic to news sites comes from Facebook, the change seems likely to bring about a critical decline in referral and reach for publishers.[37] More significant, though, was the way the new Facebook algorithm represented a profound shift in the diversity of opinions and news we see, the things that challenge us or broaden our worldview.

Sociologists describe the innate tendency to associate and connect with people similar to us as *homophily*. These similarities might be dimensions such as ethnicity, age, gender, education, political affiliation, religion and occupation or, say, where we live. We also cluster around niche interests such as whether we like pug dogs, Thai food or playing chess. The internet amplifies homophily, sorting people into online neighbourhoods on social channels like Twitter, Reddit and Facebook. It becomes much easier to find crowds of people who think, live and vote like we do online than offline.[38] It creates loud and polarizing echo chambers with less space for constructive disagreement, debate or enlightenment. As surmised by Julian Baggini, 'By retreating into bubbles of the like-minded, people can strip out a lot of inconvenient complexities a wider perspective would give, leading to a simpler but therefore also distorted network of belief.'[39]

Nearly two-thirds of Americans get news on social media, according to a recent survey by the Pew Research Center. What's more, Facebook is the number-one source of news for two in three of its users. That's nearly half of the US population.[40] The algorithm tweak means we are limiting our exposure to opposing perspectives, whether it's on a presidential race, climate change, safety of vaccinations or ISIS. It's hard for alternative viewpoints and contradictory information to break into someone's echo chamber. For the most part, we see ideas and news we are likely to agree with. Often, the cocoon is self-spun. A recent Reuters Institute Digital News Report found that 44 per cent of people in the US who use social media for news end up seeing sources from both the left and the right, at more than twice the rate of people who don't use social media. However, that's not to say they necessarily pay attention to any contrary views. When Facebook rolled out its 'related articles' feature, users continued to ignore information that undermined their favourite narrative.

If you were surprised by Trump's presidential victory or the Brexit vote, you may well be living in what Eli Pariser, author and co-founder of Upworthy, pinpointed back in 2011 as the 'filter bubble' effect.[41] 'The rise of naked partisanship, increasing economy and regional stratification, the splintering of our media into a channel for every taste – all this makes this great sorting seem natural, even inevitable,' President Obama said in his farewell speech on the evening of 10 January 2017. 'And increasingly, we become so secure in our bubbles that we accept only information, whether true or not, that fits our opinions, instead of basing our opinions on the evidence that's out there.'[42] It's no good for democracy.

In the aftermath of the EU referendum, Tom Steinberg, the British internet activist and mySociety founder, provided a powerful illustration of the filter bubble epidemic. 'I am actively searching through Facebook for people celebrating the Brexit leave victory,' he wrote. But to no avail. The algorithm must have assumed he wasn't interested. 'The filter bubble is SO strong, and extends SO far into things like Facebook's custom search that I can't find anyone who is happy despite the fact that over half the country is clearly jubilant today and despite the fact that I'm actively looking to hear what they are saying.'

People are more likely to describe 'a person like me' as the most credible source of information. A friend or, say, a Facebook friend, is now viewed as twice as credible as a government leader, according to the Edelman Trust Barometer.[43] 'The mass population is relying less on newspapers and magazines and instead chooses self-affirming online communities,' says Richard Edelman.[44] The Facebook algorithm is proof of a new 'world of self-reference'. Not only do we become victims to our own biases but it is also easier to cherry-pick stories that inflame our outrage. Distrust of institutions breeds more distrust until the fearful meme becomes contagious.

It is no coincidence that Trump pulled off a victory to become president in 2016 during a crisis of faith in traditional authorities. Here, supposedly, was a Washington 'outsider', who during his campaign promised to 'drain the swamp', clean up the political establishment. The former *Apprentice* host and impulsive falsehood-circulating tweeter, told voters he would 'shake things up' and do everything differently, from banning Muslims entering the United States to scrapping the Affordable Health Care Act (Obama Care). He promised to be the opposite of the 'very, very stupid people' currently leading America. During his campaign, Trump was consistent about one thing – he would 'Make America Great Again' (the Brexiters had a similar killer slogan: 'Take Back Control').

His claim to 'tell it like it is' represented an intoxicating form of transparency for many people. 'Those princes who do great things,' wrote Machiavelli, 'have considered keeping their word of little account, and have known how to beguile men's minds by shrewdness and cunning.'[45] In other words, Trump may have lied during his campaign, but as Stephen Colbert, the American talk-show host, explained to his viewers, Trump embodies 'truthiness': ideas which 'feel right' to many people.

'A deep recognition of the slow death of the meritocratic dream underlies the decline of trust in public institutions and the crisis of authority in which we are now mired,' says Christopher Hayes. 'Since people cannot bring themselves to disbelieve in the central premise of the American dream, they focus their ire and scepticism instead on the broken institutions it has formed.'[46] Trump's rise was a product of suffering.

At the 2016 Democratic Convention, President Obama observed that Hillary Clinton, a former First Lady, Senator and Secretary of State, was the 'most qualified candidate ever' to run for the presidency. The fact that the election was even a contest came down to trust.

I should disclose at the outset that some time ago I worked for the Clintons at their Foundation for almost three years. I respect Senator Clinton, a lot, but she personifies a breed of authority that more and more people are no longer willing to put their faith in. Senator Clinton's vote for the Iraq War; her handling of the Benghazi attack; a murky web of connections to the Foundation; her use of a private email server (and its mysterious destruction); and the revelation that on several occasions she was paid hundreds of thousands of dollars to give speeches to the same Wall Street bankers she promised to regulate – all these made her seem like a typical old-school politician and, fatally, an insider. 'Let's face it: our biggest problem here isn't Trump – it's Hillary. She is hugely unpopular – nearly 70 per cent of all voters think she is untrustworthy and dishonest,' wrote the documentary-maker Michael Moore in an incredibly prescient blog post predicting the win of Trump, twelve months before the election. 'She represents the old way of politics, not really believing in anything other than what can get you elected.'[47]

The story of Brexit is a similar tale. In a heated interview with Faisal Islam of Sky News on 3 June 2016, Michael Gove, the UK's justice secretary and leader of the campaign to leave the European Union, said, 'I think the people in this country have had enough of experts.'[48] It was a disturbing comment that really stuck in my mind. He also compared ten Nobel Prize-winning economists who signed a letter warning people about leaving the European Union to Nazi scientists loyal to Hitler who denounced physicist Albert Einstein in the 1930s.[49] They were highly controversial points to make but summed up the 'post-truth' world, a term named as the Oxford Dictionaries' word of the year for 2016. 'Relating to or denoting circumstances in which objective facts are less influential in shaping public opinion than appeals to emotion and personal belief.'[50] As Swiss-born British philosopher Alain de Botton tweeted on 15 November 2016, 'New doublespeak dictionary: elite = wretched, educated = dumb, sceptical = whining, regretful = reactionary, expert = idiot.'[51]

'Why did nobody notice it?' the Queen famously asked professors and academics at the London School of Economics during a briefing in 2008 on the turmoil of the financial crisis. Almost a decade later, it's a question being asked of experts in general, from scientists to pollsters to economists. In the run-up to the Brexit referendum, YouGov found more than half of Leave voters trusted neither academics nor economists. Two-thirds of Leave supporters – compared to just a quarter of Remainers – said it was wrong to rely too much on 'experts' and better to rely on the 'ordinary people'.[52] So why don't people trust experts? We need to believe experts are honest, have integrity and the public's best interest at heart. And sometimes we are encouraged not to trust them by vested parties because they tell 'inconvenient truths' about climate change, the economy or, say, tobacco.

Brexit and Trump are the first wave of acute symptoms emerging from one of the biggest trust shifts in history: trust and influence now lie more with individuals than they do with institutions.

'Every act of creation is first an act of destruction,' Picasso famously said. It applies to trust as much as art. As institutional trust collapses, it allows space for new systems to emerge. Technology is enabling trust across huge networks of people, organizations and intelligent machines in ways that are unbundling traditional trust hierarchies. Signs of distributed trust are appearing. Blockchain technologies, for example, which have the potential to create a digital record of history that no single person has the power to erase or change. The blockchain offers a new trust model, one where trust doesn't need to be mediated by a centralized authority such as a government or a bank but where people who might not otherwise trust each other can agree on a single truth or a common record of events.

The rise of multi-billion-dollar companies such as Airbnb and Uber, whose success depends on trust between strangers, is a clear illustration of how trust can now travel through networks and marketplaces. Tesla may look like a smart car company but in fact distributed trust underpins the grand master plan of Elon Musk, the company founder and CEO. 'You will be able to add your car to the Tesla shared fleet just by tapping a button on the Tesla phone app,' Musk has said. This will allow owners to earn money as their self-driving car picks up passengers while they are at work, on vacation or not using it for whatever reason.

It's still early days, but we have seen distributed trust powering the rise of crowdfunding sites such as Kickstarter and Patreon; social media platforms; peer-to-peer lending; open-source projects such as GitHub; Massive Open Online Courses (MOOCs); information-sharing hubs such as Stack Overflow and Wikipedia; citizen science projects; and many other person-to-person agreements and decentralized transactions that bypass traditional institutions and middlemen.

Its potential is massive but there's a catch. While distributed trust may sound like a techno-libertarian dream, the flip side is that the same tools that are being used to connect strangers all over the world can also be used in deeply unsettling and nefarious ways. Consider the profound changes in the way information and knowledge reached the public, the dark side of media abundance.

From 1962 to 1981, news anchorman Walter Cronkite was a nightly presence in millions of American homes. His kind and even-keeled voice, distinctive trimmed moustache, and calm poise, rich with gravitas, inspired deep confidence with his viewers, more than twenty million a night. Fronting CBS's *Nightly News*, he guided the country through a tumultuous period of tragedies – nuclear explosions, the civil rights movement, Watergate, the Iran hostage crisis, the Vietnam War and the death of Martin Luther King. Cronkite removing his black-framed glasses and blinking back tears to break the news of the death of John F. Kennedy became one of the defining images of the day. 'Oh boy!' he memorably exclaimed on seeing the Apollo 11 moon landing on 20 July 1969. He often embodied the emotions and thoughts of millions.

It was an era when the nightly news was a routine event day in, day out, central to many people's lives. Cronkite, who died at the age of ninety-two, was very much an old-fashioned newsman and his appeal was simple – a huge number of Americans respected, liked and trusted him. He earned a reputation as an objective straight-shooter. If Walter said it, it must be true. In 1972, he topped the rankings of the national 'trust index' poll with 73 per cent, making him the most trusted public figure in the country.[53]

'And that's the way it is . . .' Cronkite would say in his signature sign-off at the end of his broadcast. He would recite the line with humour, irony or sadness depending on the news of the day. It was a kind of definitive sign-off that newscasters couldn't get away with today – and probably never will again.

Cronkite benefited from working in a time when people had far more trust in the media and authority. During the nineteen years he was the voice of the national news, there was, obviously, no blogging. Three broadcast networks – CBS, NBC and ABC – enjoyed a monopoly; one that has now been broken into shards by the web and fragmented by millions of news producers and sources fiercely competing for views and clicks. On Facebook, the 'mainstream' news now fights for a piece of our Snapchat-sized attention spans with our friends' posts of birthday photos and pics of what they had for dinner. Cronkite earned the trust of his audiences but he also wasn't broadcasting in a 'post-truth' era. People's faith in the trustworthiness of the media has gone up in flames, a fire partly of its own making but also fanned, you could argue, by those in positions of power or influence who want to discredit a probing media. What's more, there is little consensus on who is telling the truth. The internet has made it harder to sort fact from fiction and easier to undermine the truth.

It is estimated there are more than 3 million blog posts written in the world per day.[54] One of the largest blog sites is Reddit, co-founded by Steve Huffman in 2005, when he was barely more than a teenager. Reddit is a gigantic collection of more than 853,824 message boards called 'subreddits', which range from the general – Sleep to Music to Politics to Food – to the niche.[55] For example, 'Weird Animals Without Necks', where more than 30,168 neckless-animal enthusiasts post snapshots of, well, animals that appear to be neckless.[56] Reddit also features some of the most misogynistic, racist, anti-Semitic and hateful content on the internet, including the subreddits Pics of Dead Kids, Date Rape, Jail Bait, Beating Women, Watch Niggers Die and Fat People Hate – all of which Reddit eventually banned, but it took more than two years for that to happen.[57] Reddit's claim to fame is that it is a democratic web forum for free speech and niche beliefs; anyone can say almost anything, even if it is outrageous, toxic or false.

Huffman started Reddit with Alexis Ohanian, his college buddy from the University of Virginia. The goal was simple: to become the 'front page of the internet'. The site is designed so that users 'upvote' items they think are valuable and 'downvote' those deemed to be unworthy, meaning the crowd curates what appears on Reddit's front page. It's a powerful meme engine, where stories take root and go viral. Huffman has almost reached his front-page dream – today Reddit has become the seventh most popular website in the United States, with more than a quarter of a billion unique visitors each month.[58] Reddit is huge.

'Spez' is how Huffman, thirty-three, is known on Reddit. He has spent years as a troll on the internet and doesn't try to hide that he can hack sites and is known for fake messaging his friends' girlfriends. In November 2016, he used his skills and editing privileges to tamper secretly with Reddit users' comments in the pro-Trump subreddit /r/The_Donald from 'fuck u/spez' (/r/The_Donald became the unofficial online home of the Trump campaign where an AMA – 'Ask Me Anything' – thread was used by Trump). Huffman redirected the abuse targeted at him to the moderators of the thread, without, he thought, leaving a trace of his tampering. But quickly realizing it was not a smart move to 'troll the trolls', as he put it, he fixed the comments back to their original state within the hour.

Not soon enough. The deceit, as is often the case on the internet, was rapidly discovered. It could have been viewed as simply a naive move from a young entrepreneur under pressure. Instead, the reaction was loud and intense – the meddling was regarded as akin to censorship. A couple of days after the incident, Huffman posted an apology to the community, called 'TIFU' (an acronym for 'Today I Fucked Up'). 'I am sorry: I am sorry for compromising the trust you all have in Reddit,' he wrote. 'I honestly thought I might find some common ground with that community [the trolls] by meeting them on their level. It did not go as planned.'[59]

Secretly editing another user's threads was, by Reddit standards, a gross ethical violation of power, which eroded the trust of its vast online community. Users felt betrayed that Huffman had abused his gatekeeping privileges. Lucas Schlessinger, a user from Vancouver, wrote: 'Terrible abuse of power. Makes me think what other comments have been edited by Reddit admins.'[60] The issue was not just the act in itself but the possibilities it represented.

On Change.org thousands of users signed a digital call-to-arms, a petition calling for Huffman to resign as CEO.[61] It seems, however, that Huffman, like the bankers and politicians, won't face any real repercussions for his behaviour. He merely got a slap on the wrist, just a very public one.

Huffman's abuse of power may have been short and contained but it represents something profound. We stand on the threshold of a chaotic and confusing period; a murky grey zone where institutional trust is being systemically undermined and distributed trust, for better or worse, is rising to take its place. As we overturn traditional institutions and old sources of authority, a new era of hyper-individual accountability has to take hold; one where we understand the factors that come into play when traditional gatekeepers, referees, experts and authorities are sidestepped, undermined or removed. It calls for a new kind of vigilance and decision-making. The sheer scale of the changed system presents immense challenges. For example, which assaults on trust do we choose to challenge and pursue? Why Huffman's behaviour, while a million other dubious blogs and trolls pass as good coin? It's a fallacy to believe we can take power out of individual hands. Instead, we need to think more deeply about the consequences of individual acts – and where responsibility ultimately lies.

3
Strangely Familiar

My parents, for the most part, didn't believe in cocooning their children in cotton wool. They wanted my brother and me to learn how to navigate our world through first-hand experience, by boldness and trial and error, even if it hurt a little. If we fell from a climbing frame, off a bike, a wall or even a horse, there wasn't a collective gasp. We were one of those families where minor calamities were greeted with a no-nonsense, 'Straight back up.'

But there were certain things, dangerous things, my parents taught us to avoid. Don't put your fingers in the plug socket; don't answer the front door to people you don't know; don't put plastic bags over your face; don't touch the fire or boiling kettle; don't cross the road before looking both ways; and so on.

I vividly remember the day my mum asked me, 'Do you know what a stranger is?' I was four and a half and about to start kindergarten, about to be out in the world in a new way. We were on our way home from getting my first school uniform, an awful olive green dress with a chequered green and white shirt, complete with green woollen knickers. As we walked down the high street, she made a point of explaining that a stranger is not necessarily a good or bad person, just someone we don't know. There were 'safe strangers' that were okay to trust. For example, police officers, lollipop ladies and firefighters.

And then there were others. Strangers in cars. My mother repeatedly explained that if someone stopped a car and told me to get in, I was to run and shout loudly. I was good at that. If someone unexpected was going to pick me up from school, even an aunty or Mum's friend, we had a code word to share, 'green tomato'.

Compared to the rest of her somewhat rational parenting, her fear around strangers kidnapping me in their cars bordered on paranoia.

So there is a beautiful irony that, as an adult, my work focuses on ideas that require trust between strangers, and even strangers in cars. It's a once unthinkable form of trust that has sprung up around the world. And we're only just beginning to understand how it works.

Anshul Shuka is a twenty-nine-year-old doctor who lives and works in Gurgaon, a fast-growing city on the fringes of southwest New Delhi. He regularly takes trips to Jaipur to visit his family and friends. It's a long journey of around 240 kilometres, and depending on the traffic it can take him close to four hours in his dark grey Hyundai. Rather than make the drive alone, Shuka offers his three empty seats to people who want to take a similar journey around the same time. He 'sells' the seats for 600 rupees (approx. £7.00) per passenger, advertising them on BlaBlaCar, the world's largest long-distance ride-sharing platform.

On his profile, Shuka describes himself as 'BlaBla'. It means he is a good match for other passengers who like some conversation but do not want to chat the entire journey. If he did want to talk non-stop, he would be a 'BlaBlaBla'. If his preference were to drive in silence, he would just be a 'Bla'. His profile shows that he likes listening to music, doesn't smoke and won't bring any dogs or other pets along for the ride. His rating is good, a 4.7/5. Manisha Vasdey, a twenty-six-year-old female passenger who has shared a ride with Shuka, gave him an 'outstanding' review. 'The ride was very nice and comfortable,' she commented. 'Great conversation from sports to *Game of Thrones* (which he doesn't watch!) to politics. Would definitely recommend travelling with him.'

The idea behind BlaBlaCar, a French start-up founded in 2006, is relatively simple: drivers 'sell' the empty seats in their cars on trips they are planning to take but they can only charge prices that cover petrol and road tolls. They are not allowed to make a profit (it's against the rules). It's a win-win; passengers get a relatively cheap ride and drivers offset the costs of their journey. BlaBlaCar sounds like a worthwhile community service. It is, in one sense, but it's also a commercial venture. Indeed, the company was valued at £1.2 billion in 2017.[1] BlaBlaCar makes a nice profit by charging a booking fee, which is about 15 per cent of the cost of the ride.

Notably, BlaBlaCar is different from Uber in its pricing model but it's also not designed to compete with taxis. The average trip taken is long, around 320 kilometres (200 miles), which makes it more of a direct threat to coaches and trains.[2] It is also a long time to spend in a car with someone we have never met before. We could get a back-seat driver telling us what to do on the road, someone with a weird sense of humour or, as Mum still likes to point out, a psycho-killer.

Forty-one-year-old co-founder Frédéric Mazzella first got the idea of creating a marketplace for empty seats back in December 2003. He was working hard in Paris at the time and had promised his family that he would join them for Christmas. They lived in the Vendée region of France, around 420 kilometres southwest of the capital. Busy with his job, Mazzella had left his travel plans to the very last minute. All the trains were fully booked and he didn't own a car. In a fix, he convinced his sister Lucie to make a two-hour detour to pick him up.

On that wintery night, driving along the motorway, he noticed that most of the cars had only one person – a driver and no passengers. 'I thought, that's crazy. What a waste,' says Mazzella. 'Why don't we put those empty seats in some kind of search engine, making it as easy for passengers to search and book them as it is for available seats on trains?'[3]

To Mazzella's surprise, no such website existed yet. Yes, people posted random trips on Craigslist, but that was about it. He couldn't get the idea out of his head, although for a long time he was too busy to do anything about it. Finally, three years later, in 2006, he decided the time was right to start a ride-sharing company. Online marketplaces such as Etsy and eBay that match buyers and sellers were taking off, and social networks such as YouTube and

Facebook were beginning to gather momentum. If people were starting to share photos, music and daily thoughts, why not seats in cars?

Mazzella contacted Francis Nappez, a close friend and programmer. Together they created a website that they initially called CoVoiturage (French for the term 'car-pooling').[4] The first site was very basic and ugly-looking. It didn't have user profiles and there were no peer reviews or ratings. Drivers simply submitted their emails and phone numbers and advertised their rides in a similar way to online classified ads. People had to contact one another, agree on prices and make necessary arrangements. There was a lot of friction. It took time and effort. And, most importantly, strangers were still strangers.

On paper, ride-sharing looked like a big opportunity. In France alone, there are approximately 1 billion seats travelling empty between major cities each year, and there are more than 700 million trips taken each year that are in excess of 150 kilometres.[5]

Yet the idea did not initially take off as Mazzella had envisioned. Indeed, it would take more than a decade for the platform to get real traction. People were simply not ready to take the trust leap needed to adopt this new way of travelling. That early model had overlooked something vital, something very human: the company had focused on solving the coordination problem of matching drivers and passengers but it had failed to solve the problem of trust.

In 2007, Mazzella decided to enrol in the MBA programme at the INSEAD Business School. It was there he met Nicolas Brusson, now the COO of the company. They decided to enter CoVoiturage into the INSEAD's business venture competition.[6] Disappointingly, they finished fourth. The biggest question the judges had was around trust – how was this different from a marketplace for hitchhiking?

'The iPhone did not exist. People were only just beginning to grasp the fact that we were entering the digital age. And there I am, pitching my idea of a world where anyone looking to travel could connect with drivers heading the same way,' Mazzella says. 'It took some time before the world was ready for this idea.' He had to figure out a way to get round that lesson drummed into us as children: never get into a car with a stranger.

Getting people who haven't met to trust each other to share a ride is a fiendish problem. It's basically reinventing the hitchhiking experience into one that is paid for, planned and trusted. But the problem does not in fact start with personal trust, which is getting people to trust one another. The problem starts with building *generalized trust*, which is how we first gain trust in the *idea* itself.

And for many years BlaBlaCar, as it was renamed, took the wrong approach to getting people to adopt the idea. The company tried emphasizing the environmental savings, how good sharing a car was for the planet. But it didn't work. It wasn't a good enough reason to get people to use the service. The next approach was to try to sell the software to companies who could offer car-pooling services for employees. It was a logical idea; employees are not total strangers, the trust is already there. But all the companies wanted different features to suit their specific needs. 'An enormous amount of time, resources and attention was spent on delivering multiple customized platforms. There was no scalable solution,' says Mazzella. 'Over time, it became clear the business-to-business version would not flourish.'

The turning point came when he recognized a now-obvious problem – users were frequently cancelling reservations without being penalized. 'Passengers would call, like, an hour before or three hours before saying, "Oh, I'm sorry, I can't come. My grandma is sick." We had a percentage of sick grandmothers which was way above average,' Mazzella jokes.[7] Passengers were booking seats online in multiple cars to hedge the risk of a driver cancelling last minute. And drivers compensated for unreliable passengers not showing up by overbooking the spots they had available. There was no mutual commitment and the whole experience was an inefficient mess.

So BlaBlaCar implemented a solution that in retrospect sounds ridiculously simple. In 2011, they introduced a feature so that people had to pay online and, critically, in advance. At the time of the booking, passengers were charged for the ride. Not only did this remove the social awkwardness of exchanging cash in the car, it created upfront commitment. Cancellation rates fell from 35 per cent to less than 3 per cent. And the service started really to take off. Online payments removed a *trust blocker*, things that can get in the way or are the deal breakers when it comes to people trusting a new idea or each other.

BlaBlaCar now transports more than 12 million people per quarter across twenty-two countries, as of April 2017.[8] To put that in context, it carries more passengers each month than Eurostar (2.5 million travellers per quarter) or British Airways (10 million passengers per quarter).[9] 'Nothing is as powerful as an idea whose time has come,' says Mazzella, quoting the French novelist, poet and playwright Victor Hugo.

BlaBlaCar is an illustration of how technology is enabling millions of people to take a trust leap in an idea, to do something new or different from the way we have previously done it, regardless of cultural norms.

Over the past decade, I have been researching hundreds of networks, marketplaces and systems that reinvent the way something of value – a product, service or information – reaches many people. There are fascinating nuances in how trust works in these examples and we will explore those nuances throughout the book. But beneath the differences lies a common behavioural pattern people follow in forming trust. I call it 'Climbing the *Trust Stack*'.

The trust stack goes like this: first, we have to trust the idea; then the company; and, finally, we have to trust the other person (or in some instances a machine or robot).

Let's use BlaBlaCar for an overview of how it works. On the first level, we have to trust that the *idea* of ride-sharing is safe and worth trying. There has to be enough understanding and certainty, or reduced uncertainty, to make us willing to try the idea. The next

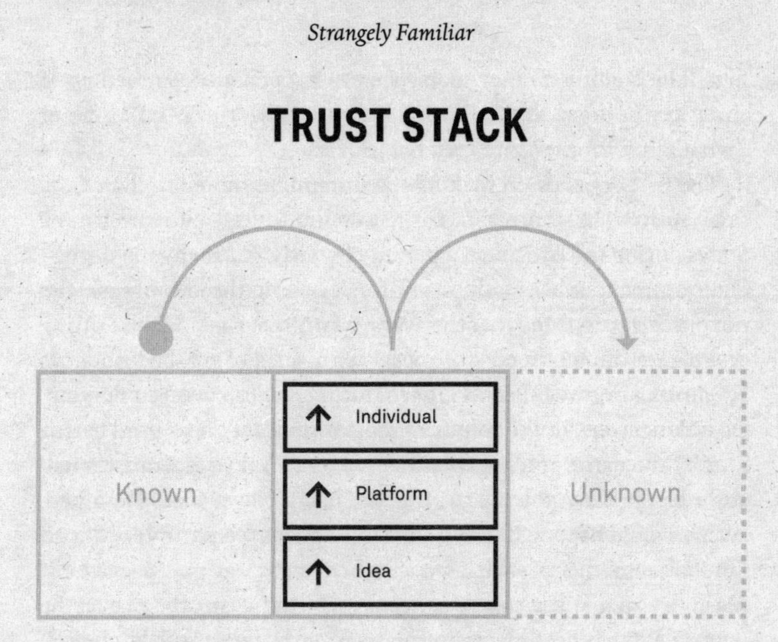

stage is about having confidence in the platform and company. In this instance, it's knowing that BlaBlaCar will remove bad apples before the ride and help us out if something does go wrong. The third and final stage sees us using different bits of information to decide whether the other person is trustworthy. It's this last level where the real trust happens. But we can't get there without going through the other two stages.

The first time we climb the trust stack it feels a bit weird, even risky. But we get to a point where these new ideas seem not only normal but necessary. We're comfortable to make the trust leap and, once made, our future behaviours change, often quite quickly.

What coaxes us into trusting new ideas? It is trust, after all, that influences how far, how fast and how permanently a new idea will spread.

Trust in new inventions doesn't happen by accident. There are some universal psychological and emotional hurdles to overcome first. The conditions that enable that to happen are summed up in three key notions: the California Roll principle, the WIIFM factor ('what's in it for me') and *trust influencers*.

The first depends on making the unfamiliar more familiar. Consider sushi. The concept of sushi was introduced into the United States during the late 1960s, a period of whirlwind change in tastes – entertainment, music, fashion and food. At first, the idea of sushi did not bite. Keep in mind that the average family at the time was sitting down to a dinner of cuts of meats with sides of mashed potatoes swimming in gravy. The thought of eating raw fish was bewildering, even dangerous, in the minds of most restaurant-goers. And then a chef by the name of Ichiro Mashita, who ran Tokyo Kaikan, a small sushi bar in downtown Los Angeles, had a clever idea. He asked, 'What would happen if the strange ingredients were combined with familiar ingredients such as cucumber, crabmeat and avocado?'[10] Mashita also realized that Americans preferred seeing the rice on the outside and seaweed paper in the interior. In other words, the roll would feel more familiar if it were made 'inside-out'.

Demand exploded. The California Roll was a gateway for many people to discover Japanese cuisine. Americans now consume $2.25-billion-worth of sushi annually.[11] As Nir Eyal, the author of *Hooked*, writes, 'The lesson of the California Roll is simple – people don't want something truly new, they want the familiar done differently.'

The California Roll principle is based on the underlying rule of combining something new with something familiar to make it 'strangely familiar'. It's a phenomenon that psychologists like the late Professor Robert B. Zajonc have labelled the 'mere-exposure effect' or the 'Law of Familiarity'. Humans, understandably, have a tendency to be more comfortable around people or like things more when they're familiar to us. In other words, familiarity can breed a kind of warmth or affection. There is more than one way to build on this.

Apple does it through a design feature Steve Jobs called 'skeuomorphism'.[12] It's a catch-all term for when design cues are taken from common objects or elements in the physical world. The iPhone calendar resembles a physical calendar. The notes app looks like a yellow legal pad. The rubbish bin on the first Mac was exactly like a metal bin. The podcast app when first launched looked like an ancient reel-to-reel tape and iBooks looked like a real bookshelf with wood veneers. The familiar elements are not necessary for the new features to function but they tap into our memory bank. Their role is to enable our brains, in a split second, to grasp things they have never experienced before. Jony Ive, the legendary Apple designer, describes the goal as 'to build things that are strangely familiar'.

When I was at Oxford doing my undergraduate degree in Fine Art, I had to take a course in modern philosophy and critical theory. We studied the likes of Descartes, Voltaire and Rousseau. The lectures would be held in a grand hall in an old Oxford building. There was lots of dark wood panelling, with enormous portrait oil paintings hung on the walls of mostly elderly men. Our professors would tell us things like, 'Art is a form of thinking that addresses the thoughtfulness of life.' I honestly thought these lectures were pompous and painful. In fact, I still have dreams about having to sit through them. At the time, I didn't know how to think in abstract ways. But there was one class that I found fascinating: Immanuel Kant. 'Human reason is by nature architectonic,' Kant wrote in *The Critique of Pure Reason* – i.e. our thoughts follow a clearly organized and defined structure.[13] Well before Steve Jobs came up with the idea, Kant believed that people need some kind of system or familiar schema to pave the way to understanding something new.

Perhaps you have stood looking at a piece of art and thought, 'What on earth is this?' I remember standing in the Tate Modern some time ago in front of a pile of 120 bricks arranged in a rectangle; bricks that would otherwise be used to make an ordinary house. The piece, by the sculptor Carl Andre, was called *Equivalent VIII*. Andre is renowned for making 'idea art' that is about 'recognizable things'. But what was the idea in bricks? I felt lost. So what did I do? I just walked away, on to the next piece. The same thing happens with new inventions and new experiences we just don't get. We move on.

It was only recently I realized that that is exactly how I think about trust. Specifically, for us to trust a new idea, we need bridges that are easy to find and to cross. The unknown needs to be reduced just enough that our mind goes, 'I get this. It's kind of like . . .' We have to turn *Equivalent VIII* into 'Destruction of the Berlin Wall'. We have to put the strange seaweed on the inside and rice on the outside.

Let me give you an example of a company that knows how to create bridges that allow people to step lightly from the known to the unknown. That company is Airbnb.

Judd Antin's official title is director of research at Airbnb. His job is to get inside the heads of guests and hosts using the accommodation platform. He wants to find out what they really think and what *really* happens. If you wanted to know, say, all the strange and unexpected ways hosts get their homes ready for guests, Antin's experience research team could tell you. Like the host who decided they did not want their guests to use a specific toilet in one of their bathrooms, so they placed a giant prickly cactus in a heavy concrete pot on the toilet seat.

Antin, thirty-eight, is a social psychologist with a PhD in information management and systems from the University of California, Berkeley. Much of his research has been on why people behave in certain ways in different online environments. For instance, how do people cope with managing multiple digital identities? Do women and men edit differently on Wikipedia? (Turns out they do.) Antin gives you the sense that when he gets interested in something, he gets a little obsessed. He is the type of person who would say no question is a dumb question, but perhaps you could ask the question differently. In fact, he said that to me a few times when we spoke over Skype.

He never talks about the people he studies as 'subjects' or 'participants'. He doesn't even use the word 'users'. It is always 'guests' and 'hosts'. He doesn't talk about 'spaces' but 'homes'. The word he uses to describe his research is 'sexy'. 'When I say sexy, I mean that the topic of our research is really meaty. The idea of letting another person into your home has lots of dimensions that you could spend years diving into,' Antin explains. 'It's research that requires deeply understanding a problem and figuring out how to turn the findings into design, product and communication solutions.'[14]

Antin's team must understand what makes people comfortable with trusting the idea of Airbnb. When guests go to the Airbnb site for the first time, they know they want to go on holiday or simply need a place to stay. But there are lots of 'what ifs' and unknowns before they get started. The first and most basic question is, 'What is the idea of home-sharing?'

Even investors had trouble getting their heads around the idea initially. In fact, many laughed the founders out of the room and dismissed it as dangerous. Take Chris Sacca, an accomplished entrepreneur and the head of Lowercase Capital, one of the leading venture funds in the United States. Sacca spots things early. I mean, really early. He has been one of the first to back the likes of Twitter, Uber and Instagram. He was one of the first people to whom Brian Chesky, Nathan Blecharczyk and Joe Gebbia, the three founders, showed the original pages of Airbnb to in 2008. But he passed on making a seed investment.

'I pulled them aside and said, guys, this is super dangerous,' Sacca said. 'You're renting out a room in somebody's house while they're still there? Somebody's going to get raped or murdered, and the blood is gonna be on your hands. There's no way this'll succeed.'[15]

Investors will run through 'what if' scenarios when assessing a deal. Specifically, 'What's the worst that can happen when using this product or service?' In the case of Airbnb, a lot can happen and has since happened.

Sacca is not the only investor who initially spurned Airbnb and missed out. I met the founders in 2009 when I was writing my first book on the so-called 'sharing economy'. I came home and told my husband, Chris, we should make an early investment. I explained in immense detail how people were going to take pictures of their homes – bedrooms, kitchens and even their bathrooms – and then guests from all over the world would book these places. He is used to me continually sharing and sounding ideas off him. But this time, he looked at me strangely, as if I had lost the plot. I went on trying to convince him: 'eBay is a multi-billion-dollar business where strangers trade all kinds of stuff including second-hand cars online. At the time, people thought Pierre Omidyar was also crazy and it wouldn't work,' I insisted. He did not budge, so I went with my overused line: 'The future is created by optimists, not pessimists.'

Chris is a barrister and therefore a master at arguing his point. 'Well, this is not eBay. People are not trading goods anonymously online. This is people's homes. This is people meeting up in the real world,' was his comeback. He was right and very, very wrong. Airbnb is now the second most valuable hospitality brand in the world, estimated to be worth $31 billion. I keep the valuation chart of Airbnb on our fridge. It has scribbled at the top, 'Always listen to your wife!'

In retrospect, it is easy to point to the likes of Sacca and my husband and say how wrong they were. But, as I say, it's also easy to see why they thought the way they did. Airbnb's success rests on having been able to get people to make that staggering leap of trust and overcome our natural 'stranger danger' bias.

Try this quick experiment yourself. The next time you are sitting next to someone you don't know, ask them to swap phones with you for just one minute. Explain that you will hold their phone and they can hold yours. Tell them: 'What you decide to do with it is your choice.'

I have played this game many different times with different groups of people, from financial advisors to students to estate agents. I have tested it at dinner parties, conference events and in the classroom. The reactions are predictable. Some people outright refuse. People laugh nervously. People hesitantly take the phone but place it face down. People ask how long is left to go. A few launch straight into the experiment, looking at messages, photos and Twitter feeds. Some even tweet or post on Instagram. But for the most part, it feels very uncomfortable for participants. And that is just holding another person's phone for less than a minute.[16]

Airbnb has had to create trust around an exchange that involves one of the most intimate things in our lives, the place where we rest our heads. 'We have to enable Olympic levels of trust between people who have never met,' Antin says.[17] And that's what makes his job so fascinating and challenging. 'As big as we are, most people will have no idea and many more people will have a vague idea of what Airbnb is, but don't really get it,' Antin told me. The first thing that will run through their minds is, 'What on earth is this anyway?'[18]

When some people go to the site for the first time, they even wonder if Airbnb owns all the homes they see on the website and the company simply rents them out. 'It sounds silly but it's not; it's a known model of holidaying they understand,' says Antin.

A lot of research has been done on how people 'get' the concept. Noticeably, there are no 'How does Airbnb work?' videos on the homepage. Admittedly, some people look at explicit things such as the 'About' or 'Trust and Safety' pages. But those are listed right at the bottom of the page. Front and centre, the first thing we see on Airbnb is a simple question, intentionally curious: 'Where?'

'One of the ways people get the concept is by relating to something they understand,' Antin told me. 'What we observe when new guests come to the site is that they don't typically go to the educational materials. Either they don't see them or they simply don't resonate. Instead, they go straight to the search box and they search for places in their hometown because it's a place they know, right?'

For example, when a first-time guest living in London wants to stay in New York, they might not search for New York but for London instead. And then they get even closer to home and search their borough, say, Camden. 'They look at the map of results and the guest's reaction is, "Oh, oh, I see. This is somebody whose house is just near mine, over there by the river, and you could stay there if you wanted to. Now I get it,"' explains Antin. 'That's the "ah-huh" moment.' Critically, Airbnb has designed the site in a way to encourage this behaviour. It could have had a drop-down menu where guests select a destination from a list. But that would be overly prescriptive. It wouldn't allow new users easily to discover something they can understand; that other nearby homes, like theirs, are available for rent.

In other words, we trust what we know but we can also trust what we think we know: ideas that are in fact quite new but appear strangely familiar.

Familiarity is not the only thing that matters in convincing us to trust a new idea. Once we are over the 'I get this' hump – the California Roll principle – the next barrier to be crossed is the 'What's in it for me?' (WIIFM) factor.

On 14 May 1796, Edward Jenner, an English doctor, carried out an experiment on eight-year-old James Phipps, the son of his gardener. After making two small scratches on the boy's arms, Jenner rubbed in a small amount of fluid from a cowpox blister. Young Phipps got the expected reaction of a slight fever but within a few days he had fully recovered. Two months later, Jenner inoculated the boy again. This time he used matter from a smallpox lesion. As the doctor predicted, and no doubt to his relief, the boy did not contract smallpox. Further tests showed that James was immune to the disease. This is how the concept of vaccinations, named after the Latin *vacca* for cow, was born.[19] So where did Jenner get the idea?

Jenner had received his training in the Gloucestershire countryside where most of his patients worked on farms with cattle. While still a medical student, Jenner observed that milkmaids who contracted cowpox, a relatively mild disease, did not catch smallpox, one of the deadliest infectious diseases of the period, claiming the lives of millions of people including five reigning monarchs. A cure was urgently needed.

In 1797, Jenner sent a short paper to the Royal Society describing the findings of his experiment. The paper was rejected. Scorned by his peers, Jenner published the book himself, documenting his theory that cowpox did indeed protect against smallpox. His ideas, however, were met with widespread controversy and criticism. The reasons varied from religious to scientific and political objections. The clergy claimed it was repulsive and 'unchristian' to inoculate someone with material originating from an infected cow, one of God's lowlier creatures. Some physicians dismissed his research as 'unethical'. Others selfishly did not want Jenner to succeed because they were making significant sums of money selling different kinds of drugs to treat the disease. In 1802, a satirical cartoon was published called 'The Cow-Pock – or – the Wonderful Effects of the New Inoculation!', summing up the sentiment of the time. It caricatured a scene of the vaccine being administered to a group of petrified young people and cows sprouting from different parts of their bodies.[20]

Jenner was told that his ideas were too revolutionary and that he needed more proof. Undaunted by the ridicule, he experimented on several other children, including his own eleven-month-old son, by placing a small amount of cow pox scab into the human skin. Every child he injected showed immunity to the disease.

The young country doctor did not discover the idea of immunization – it had a long history in China and Africa – but he was the first person scientifically to attempt to control an infectious disease by the deliberate use of vaccination. Jenner later became known as the 'Father of Immunology'. In 1853, thirty years after Jenner's death, the British government made cowpox vaccinations compulsory in England and Wales. On 8 May 1980, the World Health Assembly declared that the world was free of smallpox.[21] But for decades Jenner's new methods did not catch on. Why? People were not ready to take a trust leap in the idea of vaccinations because they didn't fully understand the risks and benefits – they couldn't see what was in it for them.

And it's a doubt that persists today. The anti-vaccination movement has existed since Jenner first discovered the cure, but it gained resurgence during the late 1990s after a series of events stoked the fear and doubt already brewing. One was an article that appeared in the UK medical journal *The Lancet* in 1998, written by Dr Andrew Wakefield and his colleagues, implying a possible link between the measles, mumps and rubella (MMR) vaccine and autism. Remarkably, Wakefield wrote the paper based on research he had done on a mere twelve children.[22] The children in the study were carefully selected and the research was largely funded by lawyers acting for parents who were involved in lawsuits against vaccine manufacturers. The article has since been discredited and retracted but the damage was already done. Dr Sharon Kaufman, the chair of the Department of Social Medicine at the University of California, has done extensive research focused on individuals' trust – and mistrust – in the findings of medical knowledge. 'Most parents are not anti-vaccination *per se*,' she says. 'Rather, they live in a time, as we all do, of heightened risk awareness, mistrust of government institutions and the pharmaceutical industry and a great many opportunities for "seeing" doubt.'[23] Even if people don't believe that vaccines cause harmful side-effects such as autism, the power of anti-vaccination stories is they evoke fearful emotions that undermine our trust, no matter what we might rationally think.

James Samuel Coleman, a famous American sociologist born in 1926, was, among other things, fascinated by how we make decisions.[24] Specifically, by the ways people decide whether or not to trust a new idea. He was living in a time when many great new technologies, from the first television and video recorder to the first commercial passenger jet plane, were introduced. Essentially, his research showed that we decide whether to trust based on assessment of the upsides and downsides. We make a calculation about whether trusting this idea will in some way make our lives better or not.

It sounds obvious, right? But it raises a critical point – we don't *want* to use a new invention until we *understand* it. That doesn't mean we all need to understand precisely how a technology functions, whether it was a combustion engine back then or a blockchain now. However, we do need to grasp what it can do and what it can give us. Until that chasm is crossed, we won't abandon what we already have.[25]

Creating trust in a television is one thing. What about creating it in something that has the power to hurt or even kill us? We're seeing this play out now with self-driving cars. Autonomous vehicles are expected by some engineering groups to account for up to 75 per cent of vehicles on the road by 2040.[26] But how do you get people to trust machines enough to let them take over the wheel? An expert who knows the issue first-hand is Dr Brian Lathrop.

Lathrop has worked in the Electronics Research Lab at Volkswagen (VW) since 2004. He has a PhD in cognitive psychology, specializing in human interface design, and is the person in charge of research and development of VW's autonomous vehicles.

Prior to his current role at the automaker, he worked at NASA during the period when pilots were first adopting plane automation. He became fascinated by a rather worrying state he calls 'mode confusion'. 'This happens when pilots wonder if the aeroplane is doing the flying or if they are,' Lathrop explains to me. 'Autonomous vehicles face exactly the same challenge.'[27]

During our conversation, Lathrop speaks surprisingly openly about the challenges of self-driving vehicles. He has a natural exuberance but you wouldn't describe him as an evangelist. More like an optimistic realist.

In March 2016, the American Automobile Association (AAA) conducted an extensive survey to understand how much trust its members had in self-driving cars. Three out of four drivers in the United States said they would feel 'afraid' to ride in self-driving cars. Only one in five said they would trust a driverless vehicle to drive itself with them inside.[28] The reasons people gave included: 'trusting their own driving skills more than the technology' (84 per cent); 'feeling the technology is too new and unproven' (60 per cent); and 'not knowing enough about the technology' (50 per cent).

There is a YouTube video of a seventy-year-old grandmother,

named Shirley, freaking out the first time she is inside a Tesla Model S on autopilot. As the car moves across lanes and navigates traffic, Shirley screams, 'Oh no, there is a car coming! Oh dear Jesus. Ah, ah, ooh, ooh, where is it going?' It is painful to watch. In the video, you can hear her son Bill laughing at her reaction after he turns on the autopilot. 'Oh my, I am about to die,' his mum cries out. She looks like she is on the verge of heart failure.[29] In fact, all Shirley needs to do to take control from the machine is simply touch the steering wheel. But her mind is elsewhere.

When I ask Lathrop how hard it is to get people to trust being driven by autonomous vehicles, I assume he will go into details of how smart design can overcome the fears of people like Shirley. I expect him to cite the safety stats. In fact, he makes a quite different observation. 'People trust the car quickly, almost too easily,' he says.

Yes, Lathrop has seen how some people, such as Shirley, freak out the first time they are driven by an autonomous car. They are, however, the minority. Others are completely awestruck and think, 'Wow, it's doing the driving for me.' But then something interesting happens. After a few miles – around twenty minutes of being driven by the car – the experience feels normal, even boring. Being driven by an intelligent machine it turns out is just not that exciting. That's why Lathrop is worried about people nodding off.

'What do you do when people fall asleep in a self-driving car?' is his biggest concern. 'How do you deal with that?' He thinks it's a question that is not being asked enough. 'It's not good for the trust equation!' Lathrop admits half jokingly.

The idea of people trusting the car too easily is intriguing so I press Lathrop harder. Surely, the first step in getting people to use self-driving cars has to require a massive trust leap? 'Think about it,' he replies. 'Most people are comfortable being a passenger. They are used to being driven by a colleague or by a friend.' In other words, the idea of being a passenger in a self-driving car taps into something familiar. 'The leap we are asking people to take is not a new experience, but it's asking people to trust a machine versus a human to drive. It's not that huge.' It is strangely familiar.

Lathrop admits that before people get inside an autonomous vehicle, they do have questions. Quite predictable ones. 'What if a vehicle cuts in front; will the car respond?' 'Can the vehicle change lanes?' And the most common is, 'Can it drive as well as me?' Lathrop points out that very few of us ask about unpredictable situations. For instance, what happens if a deer or even a dog runs out in front of the car? How does the car navigate parking lots at busy sports matches? The researchers have not yet figured out the answers to all these questions. Some are hard to simulate safely or even predict. It's those unforeseen, or as Lathrop puts it, '1 per cent corner cases', that his team needs to solve for. But the point is, very quickly a first-time user runs out of questions. They just get into the vehicle and let it do the driving.

The ultimate success of self-driving cars – that it becomes normal to use one – doesn't depend on engineering success. It doesn't even depend on us understanding how the technology works. It depends on that second principle of getting people to trust an idea, the WIIFM factor. We want to know what we will gain. In this instance, will the benefits of the machine doing the job of a human outweigh the risks?

The typical American commuter spends on average more than fifty-two minutes per day stuck in traffic. That adds up to more than 4 billion hours of wasted time in the United States alone, time we could use in better ways.[30] 'People want to know what they will be able to do if they are freed from driving the car,' Lathrop tells me. They imagine being able to watch movies, talk on the phone, work and eat. 'Look around you when you're at the traffic lights, when cars are stopped. These behaviours are not new. I want to allow people to do what they are already doing but in a much safer manner,' he says. Indeed, the biggest benefit of self-driving cars is safety.

One hour after passing my driving test as a teenager, I crashed my car. And it wasn't a minor ding. I was waiting at the intersection of a busy crossroads in Hampstead Garden Suburb, north London, looking for a break in the traffic. I started fiddling with the radio, switching from Capital to Virgin and other stations, trying to find a song I liked. Distracted, I decided foolishly to make a go for it. Smash. You never forget the sound of metal hitting metal. My car spun and landed in what was previously a nicely groomed privet hedge on the other side of the road. Over the course of the next five years, I had three subsequent accidents. I am truly a terrible driver. And my experience is a reminder of why self-driving cars matter.

Human error and inconsistent driving cause more than 90 per cent of crashes, which kill more than 1.2 million people annually, according to the World Health Organization.[31] It is estimated that driverless cars could, by mid-century, reduce traffic fatalities by up to 90 per cent. It works out to be more than 300,000 lives saved each decade in the United States alone, and a saving of $190 billion each year in healthcare costs associated with accidents.[32] In the UK, KPMG estimates that self-driving cars will lead to 2,500 fewer deaths between 2014 and 2030.[33] But we are still highly sceptical of the benefits. 'No one is going to want to realize autonomous driving into the world until there's proof that it's much safer, like a factor of 100 times safer, than having a human drive,' Andrew Moore, the computer science dean at Carnegie Mellon University told *The Atlantic*.[34]

Humans are not 'risk mutual' people. In other words, they don't place the same weight on things going well as things going badly. For example, say I lost a favourite navy blue chequered coat, I would feel more about the loss than I would about the joy of finding another coat, even if the jackets were identical. That's the human tendency, to feel more strongly about a loss than a gain. The basic idea of 'loss aversion' is a concept first discovered by Daniel Kahneman and Amos Tversky, the Israeli psychologists who changed how people think about how people think. Sticking with the status quo feels much safer and better, even if we know we might compensate a loss.[35]

Undisputedly, we love new things, but those things tend to be an upgrade or improvement on ideas we are already comfortable with. For instance, the high-definition wireless flat screen is better than the clunky colour television box with wires and remote controls. It is easy to trade up. To date, the media has not done much to help foster the idea that self-driving cars are safer than humans. In fact, it has tended to fuel the opposite image: the car could kill you.

On 7 May 2016, Joshua Brown's Tesla Model S vehicle crunched into the side of an eighteen-wheeler trailer truck. Forty-year-old Brown was killed. The car was on autopilot during the collision. When police arrived at the scene, a *Harry Potter* movie was reportedly playing on a portable DVD player inside the car. In a blog post published on the day the accident became public, Tesla stated, 'Neither Autopilot nor the driver noticed the white side of the tractor trailer against a brightly lit sky, so the brake was not applied.' In other words, the car is not perfect. And the driver was distracted by a young wizard.

'The media love these stories. They sensationalize them. I understand why,' Lathrop tells me. 'But they are extremely rare.' Indeed, this was the first known autopilot death in roughly 130 million miles driven by Tesla customers.[36] To put this in perspective, among all vehicles in the United States, it is estimated there will be a fatality every 94 million miles.[37] In other words, self-driving cars are 36 million miles safer.

Lathrop thinks a lot of the messaging around self-driving cars is not useful. 'We need to temper expectations that an autonomous vehicle can be perfect,' he says. 'We need people to realize that the benefits hugely outweigh the negatives.'

What's interesting in Lathrop's observation is that it illustrates how building trust doesn't have to hinge on a promise of perfection. In fact, guaranteeing 100 per cent certainty of an outcome is a recipe for disaster.

Whether it's deciding to use Airbnb rather than a well-known hotel brand such as Marriott or Hilton, or deciding to put our trust in an intelligent vehicle, the pros and cons we draw up will circle around the same dimensions – value and certainty. Whatever the idea, the questions are essentially the same: Will these experiences create value in my life? How can I be sure of that value?

*

You've no doubt heard the term 'early adopter'. It typically refers to an individual (or business) who uses a new product or technology before others do. My brother-in-law is one. He seems to be in on the next big thing way before it's big. 'Found it first' is a badge he likes to wear. He tells you about new things you should be trying: digital wallets, the Aire digestive tracker that helps you figure out which foods are most compatible with your body, the Nest Learning Thermostat and so on. Early adopters tend to seek novel things to tinker with, and then become evangelists for them. They tend to have strong and excited opinions (and I don't mean that in a bad way). Let's just say they know a lot without being a know-it-all. Without a doubt, early adopters are critical to innovations taking off. Interestingly, however, as a group they are not necessarily the most influential in terms of getting the late majority, or laggards, to climb the first layer of the trust stack. What is needed is that third element – made up of what I call the trust influencers.

Trust influencers are groups of people who can disproportionately influence a significant change in the way we do something; they set new social norms.

Trust influencers are out there for every idea but sometimes we have to look hard and in unexpected places to find them.

There is a wonderful example of trust influencers relating to what is not a particularly exciting concept – transferring money overseas.

Behind almost every great start-up lies a story of personal frustration. The tale of TransferWise goes like this: Taavet Hinrikus was born in Estonia when it was still part of the USSR. It was a tough environment to grow up in. 'One really had to take charge and solve problems in creative ways to get anything done,' Hinrikus says.[38] In 2002, when he was twenty years old, he met two budding entrepreneurs called Niklas Zennström and Janus Friis. They were tinkering around with an idea: what if people could digitally transfer their voices and words to one another? Like a phone but without a phone. And without the phone bill. This was exactly the description first used for what would become known as Skype. Hinrikus became one of the first employees on the team.

Skype grew fast. In 2006, the company needed Hinrikus to move from Estonia to London to help them expand. He was earning euros that were paid into his Estonian bank account. Every fortnight Hinrikus had to transfer money to a UK account, to cover his rent, food and other costs. When his money eventually arrived, it was much less than he expected. The whole experience, he thought, was unnecessarily painful.

An old friend, Kristo Käärmann, faced the same problem but in the opposite direction. He was working as a consultant at Deloitte and was being paid in pounds. He was transferring money back to Estonia to pay his mortgage. 'I was losing five per cent of the money each time I moved it,' Käärmann says. The money was being lost in the hidden fees and poor foreign exchange rates that earn the banks a hefty profit.

Hinrikus and Käärmann came up with a simple but clever idea. 'We figured we could just "swap" the money. I could just transfer money from my Estonia account into his Estonia account and he transfer money from his account in London to my account in London,' says Hinrikus. 'Pretty quickly we saved thousands by eliminating unfair rates and banking fees.' The friends realized they had hit on an enormous opportunity. TransferWise, the company they founded in London in 2011, is as of March 2017 valued at more than $1.1 billion and already has a 5 per cent market share of international money transfer in the United Kingdom.[39]

TransferWise is based on a peer-to-peer technology system that matches money flows. Say I wanted to send £1,000 from a bank in London to a bank in Paris, the system looks for someone else wanting to convert euros to pounds. The result is that the money never moves between countries. Doing things this way means the process is faster, easier and cheaper than the transfer services we get from banks. Which brings us back to the question of trust and changing behaviour.

The conventional way to transfer money from one country to another is to use a traditional bank or a post office or a known brand such as Western Union. In 2015, in excess of $601 billion was transferred in this way.[40] So what would persuade ordinary people to trust an unknown digital start-up to transfer their money? The answer: seeing unexpected users putting their faith in this new way of doing things. People who make us think, 'Hey, maybe this idea isn't so risky after all.' But just who would TransferWise's trust influencers be?

Hinrikus and his team realized their ideal trust influencers were neither fintech know-it-alls, nor the people with the latest Apple Watch. Far from it. They had to find users we wouldn't necessarily expect to take a risk with an unknown company like TransferWise. Namely, pensioners. Retired British people living abroad in Spain, for instance, who needed to get their pension regularly transferred from pounds to euros. 'For them, the fees they were being charged represented a big portion of their total, so they had a strong incentive to take the same leap of faith,' Hinrikus says.

When other first-time users heard about pensioners giving TransferWise the thumbs-up, it had enormous influence on shaping their decision to trust the idea. Indeed, when enough trust influencers are seen to have made the trust leap and survived, millions will follow, often very quickly. That's how change spreads.

James Surowiecki wrote brilliantly about the influence of group example in his book *The Wisdom of Crowds*. It was largely based on the science of crowd persuasion: how groups of people can influence other individuals to say 'yes'. Some of the core ideas were grounded in Professor Robert Cialdini's theory of 'social proof'. 'If a lot of people are doing the same thing, they must know something we don't. Especially when we are uncertain, we are willing to place an enormous amount of trust in the collective knowledge of the crowd,' Cialdini wrote.[41] Put simply, we tend to follow the lead of other people, especially when we are unsure.[42]

Various experiments have shown the different dimensions of

social proof. One of the most visual is known as the 'Street Corner Experiment'.[43] It was designed in 1968 by social psychologists Stanley Milgram, Leonard Bickman and Lawrence Berkowitz. First, the researchers put a single person on a street corner and had him look at the empty sky for sixty seconds. Only a fraction of passers-by stopped to see what the person was looking at. So the next day, they put five people staring at the sky on the same street corner. Four times as many people stopped to check what they were gazing at. And when the researchers put fifteen people to stare at the empty sky, 45 per cent of all passers-by stopped and tilted their heads to see what the others were looking at. They stopped traffic by staring at an empty sky. The typical takeaway from this study, as Surowiecki puts it, is 'the crowd becomes more influential as it becomes bigger'.[44]

There's no doubt social proof builds trust around new ideas, especially when we are uncertain of the outcomes. That is why we commonly see sites boast the number of reviews or users they have. For instance, TransferWise has on its homepage, front and centre, their number of 'happy reviews' (43,168), and its number of customers (more than 3 million).[45] Indeed, we seem to be living in a world where it is perceived that the critical way to persuade us is through large numbers, be it Facebook 'likes', five-star ratings or Twitter and Instagram followers.

But social proof, and the trust it fosters, does not have to come from large crowds. It can also come from a small group of individuals with a unique power to influence. They do not need to have an impressive title, be a celebrity or even a credible 'expert'. They do not need large followings. They do not even have to be people who are similar to the majority, the crowd. They can be just like the British pensioners in Spain, people who can change other people's perceived uncertainty because they seem the least likely to take a trust leap.

The three ideas covered in this chapter – the California Roll principle, the WIIFM factor and trust influencers – aka 'What is it?', 'What do I gain?', 'Who else is doing it?' – offer a way to see how an idea once dismissed as preposterous can turn into something strangely familiar. They explain how trust in new ideas spreads. Anyone who has ever built trust in a venture, a new product or an idea has had to go through that process, whether they're aware of it or not.

It's as if there's a daunting rock face the creators of the new idea are asking people to climb. First, they have to show the would-be climbers some familiar moves and handholds, to reduce the unknowns enough to encourage that first step. (And while they've reduced uncertainty, they haven't promised perfection – there's still risk.) They have to explain what climbing the rock face has to offer. Finally, they have to point out the other climbers above them who are loving the experience. Before long, the doubters find themselves racing up the climb, leaving the ground so far behind it's soon just a distant memory.

And that process is powerful. It can turn an idea once dismissed as risky and even frightening – sharing a long ride with a stranger, staying in the home of someone you don't know or getting into a self-driving car – into something normal, rewarding and disruptive. So that's the 'idea' part of the bigger climb, the climb up the trust stack. The next level is building trust in the platform.

4
Where Does the Buck Stop?

On 20 February 2016, in Kalamazoo, Michigan, a deadly rampage played out. Over the course of five hours, a forty-five-year-old Uber driver named Jason Brian Dalton became a mass killer, shooting six people dead and leaving two seriously injured. In between the separate incidents of bloodshed, Dalton went back to routinely picking up Uber fares.

The shootings began around 5.40 p.m. EST on a Saturday evening and the first victim was twenty-five-year-old Tiana Carruthers.[1] Carruthers was crossing a parking lot with five children, including her young daughter, when a silver Chevy Equinox with Dalton at the wheel veered towards her. Dalton rolled down his window. He asked if her name was Maisie (the name of the passenger he was circling around the estate trying to find and pick up was Maci.) 'No, I am not that person,' Carruthers replied.

Something didn't seem quite right about the driver. She yelled at the children to run. Dalton sped off but then quickly turned the car round and headed directly for the terrified Carruthers. Pulling out a Glock 9mm semi-automatic pistol, he shot her at least ten times, hitting her arms and legs, the last bullet lodging in her liver. Remarkably, she survived.

Until that day, Dalton was by all accounts an ordinary Joe. Neighbours described him as 'a little odd and awkward' but generally sociable. He worked as a loss adjuster for a local insurance company. He had been married for almost twenty years and had two children: a fifteen-year-old son and a ten-year-old daughter. Dalton became an Uber driver on 25 January 2016, about a month before the shooting.[2] He wanted to earn some extra money to take his family to Disney World. Before that tragic evening, he had clocked up more than a hundred Uber trips and passengers had given him a very good rating after rides: an average of 4.73 out of a possible five stars.

On the day of the shooting, Dalton had run a few errands, including a visit to the local gun store. It wasn't unusual – he loved guns and owned sixteen firearms. He turned on his Uber app around 4 p.m. Shortly after, he picked up a young man called Matt Mellen. It seemed like a normal ride until Dalton took a call from his son, at which point something abruptly shifted in his mood. He accelerated wildly and took Mellen on a hair-raising ride. 'We were driving through medians [central reservations], through lawns, speeding along,' Mellen later told the police.[3] Dalton even side-swiped another car but seemed completely unfazed by what he had done. 'He wouldn't stop. He just kind of kept looking at me like, "Don't you want to get to your friend's house?" and I'm like, "I want to get there alive,"' said Mellen. When the car finally screeched to a halt, Mellen jumped out.

Both Mellen and a concerned bystander put in calls to 911, describing the car and Dalton. Mellen specifically identified him as an Uber driver. 'I don't want somebody to get hurt,' he told emergency services. They didn't seem too concerned. Mellen then tried getting hold of someone at Uber, to get the car off the road. Uber didn't seem to prioritize the call, even though Dalton's whereabouts could have been easily located through GPS tracking on the app.[4] It was the first of many alarm bells that would go unheeded.

Soon after that, Mellen's fiancée posted Dalton's Uber photo on Facebook with a lengthy warning: 'ATTENTION Kzoo peeps!!! This Uber driver named JASON drives a silver Chevy Equinox is NOT a safe ride!' she wrote. 'Hoping this man will be arrested or hospitalized soon if he has a medical condition causing his behaviour.' Instead, Dalton went from his crazy ride with Mellen to shooting Carruthers.

After that first shooting, Dalton swapped his car for his wife's black Chevrolet. Carole Dalton later told police he seemed a little 'troubled'; he had also told her not to go back to work, to fetch the kids, stay inside and lock all the doors. Dalton, meanwhile, went back to picking up regular Uber fares. @IamKeithBlack tweeted that he got a ride with him at 8 p.m., including a clear screenshot of the driver. He had given Dalton a five out of five star rating after his ride. When he heard about the shootings, he tweeted 'Lucky to be alive'.[5]

Others weren't so lucky. In the lot of a nearby brightly lit car dealership around 10 p.m. that night, seventeen-year-old Tyler Smith and his father, Rich, were looking at vehicles, while Tyler's girlfriend, Alexis, waited in the car. Dalton drove in, parked and walked up to the father and son, fatally shooting both of them. A terrified Alexis hid until he drove off.

Fifteen minutes later, in a final burst, Dalton killed four older women and critically injured a fourteen-year-old girl. Nothing connected the victims – male, female, white, black, young and old – and they weren't targeted for any reason.

Late Saturday night is one of the busiest periods on Uber. Remarkably, after killing six people and going home to change guns, Dalton carried on picking up revellers requesting rides. By now, the shootings were all over the news. Some passengers had heard there was a mass murderer called Dalton, an Uber driver, on the loose in Kalamazoo, yet they continued to use the app to get to where they wanted to go. A passenger named Marc Dunton even asked Dalton, 'You're not that guy going around killing people, are you?'

'Wow,' Dalton answered. 'That is crazy. No way – I'm not that guy.'

Only one passenger made the connection and refused the ride. The young woman's father had texted and called several times to warn her. She requested an Uber around 12.30 a.m. 'Jason. Chevy Equinox' came up on the phone as he was the nearest driver. She cancelled the ride. The same thing happened on her next attempt. A few minutes later, Dalton was arrested in the parking lot of a downtown Kalamazoo bar. When police officers asked the shooter to explain his motive, he replied, 'I don't think there is a why.'

During his interrogation Dalton said he recognized the Uber logo as being the religious Eastern Star symbol. He said a horned 'devil figure' would pop up on his phone through the app and would cast an intoxicating spell over him. 'It would give you an assignment and it would literally take over your whole body,' he said.[6] When asked why he randomly shot people, he calmly claimed, 'The Uber app made me do it . . . It just had a hold of me.' He has since been accused of six counts of open murder, two counts of attempted murder and eight counts of felony firearm.

The day after the shooting, Uber issued a press release. 'We are horrified and heartbroken at the senseless violence in Kalamazoo,' announced Uber's chief security officer at the time, Joe Sullivan. 'Our hearts and prayers are with the families of the victims of this devastating crime.' A few days later, the company asserted that the rampage could not have been predicted. 'There were no red flags, if you will,' said Sullivan. 'Overall his rating was good, 4.73 out of five.' Perhaps realizing that pointing to Dalton's trust rating was not the best line of defence, he later added: 'As this case shows, past behaviour doesn't always predict how people will behave.'

It's true, there may have been no reliable way to predict that Dalton would morph from Uber driver to psychotic mass shooter during a single shift. But why didn't anyone seem to respond with any urgency when Mellen first contacted Uber about an hour before the first murder? Turns out, the complaint wasn't prioritized by Uber's customer response team because it wasn't explicitly about violence. 'He said the gentleman was driving erratically,' an Uber spokesperson said, pausing. 'Remember we're doing three million rides a day. How do you prioritize that feedback and how do you think about it?'[7] Still, killings aside, you would think side-swiping a car, speeding and running up on the central reservation might raise some immediate alarm about a driver's fitness to be transporting passengers. Who could have foreseen such tragic consequences? But whose job is it to respond and act?

*

The horrific shootings intensified scrutiny on how Uber decides who is fit and safe to be a driver. The company claims it spends tens of millions every year performing background checks on applicants. San Francisco District Attorney George Gascón and many other critics have called those background checks without fingerprints 'completely worthless'.[8] Dalton had passed the checks before he started driving on the platform. But here is the problem: he had no prior criminal history, ever. He came out clean as a whistle.

Uber doesn't run the fingerprint checks that taxi and limo companies are required to do. But would they help anyway? 'As the Equal Employment Opportunity Commission (EEOC) has emphasized, background checks have limited predictive value and can have a disparate impact on minority drivers,' says Brishen Rogers, an associate professor of law at Temple University who is researching the social costs of Uber.[9] That said, fingerprinting might have at least revealed that Dalton owned sixteen guns, assuming he held a Federal Firearms License.

While Dalton's crime was by far the worst an Uber driver has committed, there has been a troubling stream of other serious incidents. Uber drivers from Boston to Los Angeles, Delhi to Sydney, have been arrested for sexual assault, rape, kidnapping, theft and drink-driving while carrying passengers. In April 2016, a driver was arrested for slashing a passenger's neck. On 23 May 2016, another was accused of strangling a student in a parking lot.[10]

In February 2017, the company was hit with sexual harassment allegations and former employees accusing the company of fostering a toxic, misogynist culture. Shortly after, a damning video surfaced online of former CEO Travis Kalanick verbally abusing a driver in the US called Fawzi Kamel, who had complained to the CEO about dropping prices and lower pay. 'People are not trusting you any more,' said Kamel. 'I lost ninety-seven thousand dollars because of you. I'm bankrupt because of you.'

'Bullshit,' Kalanick angrily retorted, telling the driver his problems were his own fault. 'Some people don't like to take responsibility for their own shit. They blame everything in their life on someone else.'[11] A fuming Kalanick then slammed the door. He later apologized for his actions.

Five days later, it was revealed that Uber had secretly used a software tool called Greyball, designed to identify city officials attempting sting operations to catch Uber drivers violating local regulations. Then Google's self-driving car outfit, Waymo, filed a lawsuit accusing Uber of stealing technical trade secrets to fuel its own self-driving car research. Following this string of scandals, Jeff Jones, Uber's company president, and six other key executives resigned. Kalanick, the notoriously hard-nosed co-founder, resigned as CEO in June 2017.

Clearly, for all its success – at its latest valuation of $72 billion it is the world's most valuable private start-up in history – Uber has had many serious breaches of trust.[12] And yet more than 5 million people every day, myself included, still tap the Uber app, and within minutes get in a car with a total stranger, often without a second thought.[13] We have, in a sense, outsourced our capacity to trust to an algorithm, and that trust, perhaps for convenience's sake, has proved hard to destroy. The question is, where does Uber's responsibility start and finish?

There are dangerous taxi drivers and, for that matter, dangerous people in any industry. Here's the difference: in Uber's terms of service, the company denies any liability for how third-party drivers – whom Uber considers to be 'independent contractors' – behave on its platform. The company says it is merely *facilitating* the needs of people who want to drive, and you are getting a ride from them in *their* car, not a company car. 'Your day belongs to you,' Uber enthuses to would-be drivers. Uber takes up to a 25 per cent cut of the total fare, a service fee, to play the role of go-between.[14] The company claims it can't control what drivers do on the job as they are interacting with an automated system delivered primarily via an app. Uber, however, is not a neutral platform, like a phone line, simply matching supply and demand. It controls surge pricing that temporarily raises fares, and drivers can be suspended for not accepting enough rides or for low passenger ratings. Uber has been involved in more than 170 lawsuits in the US alone, from class-action safety complaints to price gouging to data failures to privacy practices to the biggest lawsuits of all, the misclassification of drivers.[15]

Issues of accountability are incredibly complex in an age when platforms offer branded services without owning any assets or employing the providers. Tom Goodwin, a senior vice president at Havas Media, put it well when he wrote in an article: 'Uber, the world's largest taxi company, owns no vehicles. Facebook, the world's most popular media owner, creates no content. Alibaba, the most valuable retailer, has no inventory. And Airbnb, the world's largest accommodation provider, owns no real estate. Something interesting is happening.'[16]

When disasters such as the Kalamazoo killings happen, they raise the question of where accountability should lie when things go wrong.

It's much easier to know who to blame when traditional brands breach trust. Take the Tesco scandal that took place in January 2013.[17] More than 10 million hamburgers and other meat products were withdrawn from supermarkets after traces of horsemeat were discovered in some of its beef products. The disclosure sparked a national outcry and it became one of the biggest food scandals of the twenty-first century. Even the then prime minister got involved. David Cameron reassured the British people that everything possible would be done to address a 'very shocking crime'.[18]

In the wake of the scandal, Tesco issued an 'unreserved apology' and promised to introduce a robust new DNA-testing system to ensure the food customers buy is exactly as the label says. How did the 29 per cent horsemeat in the Tesco burger, falsely labelled 'pure beef', get in there? Although Tesco publicly accepted responsibility for the fiasco, they appeared to pin much of the blame on their supplier, Silvercrest, owned by the ABP Food group, who had 'breached the company's trust'.[19]

*

When a customer shops at Tesco, their trust clearly lies in the supermarket brand, what they experience in the store and the products they buy. Tesco, the company, has to behave in trustworthy ways so that their shoppers trust *them* and their products. But where does trust ultimately lie with platforms?

When I get in a car with a stranger, is it the driver I am trusting? Have I placed some faith in Uber, the company, its team? Am I trusting the Uber brand? Perhaps I have confidence in the platform itself, the app, payments, rating system and its mysterious pricing algorithm? Some of the answers lie in the history of trust between people, companies and brands.

There was a time when people lived in tiny communities, hamlets made up of perhaps no more than a hundred people. Everybody knew everybody else and relationships were tight-knit. People's trustworthiness, or lack of it, was evident to everyone, given the close proximity.

As hamlets turned into villages and small towns, the population tipped well above what has become known as 'Dunbar's number'. The famous University of Oxford psychologist and anthropologist found that our brains, on average, are designed to have a limited number of people, around 150, in our social group. Yes, you can friend 500, even 5,000 people on your Facebook page, but Dunbar asserts that it is hard for us to maintain stable, meaningful relationships (online and offline) beyond 150. Within that 150, an inner circle of fifteen is the very limited number of people that you turn to for support when you most need it. And this circle of meaningful relationships is fluid: the friend you confide in this week may not be the person you turn to the following month. On the flip side, our brains can handle group sizes of up to 500 at what Dunbar calls the 'acquaintance level'[20] – put simply, the people with whom we can put a name to a face.

When people moved into larger towns, with populations way above Dunbar's number, a close circle of trust, based on direct knowledge of each other, was no longer possible. Our reputations became an essential asset. If the baker offered good bread, people would buy it and others would hear about its quality. Equally, if the local blacksmith did shoddy work or someone failed to pay back a loan, word would get round. That dynamic kept most people up to the mark. The evolutionary biologist Robert Axelrod called it in his classic book *The Evolution of Cooperation* the 'shadow of the future',[21] referring to the idea that people behave better or more cooperatively when they know they're likely to meet or meet again (as opposed to a one-off encounter) and might be judged on previous behaviour.[22] These local traders knew that how they behaved in the present would shape future prospects. Just like the Maghribi traders, it's the promise of benefits from continued cooperation that helps keeps us in line.

Even the earliest incarnations of what we now call a 'brand' first rode on the back of personal reputation. The agricultural machinery giant John Deere was founded in the 1830s by a young entrepreneurial blacksmith living in Illinois who had invented an innovative plough. The Mars brand empire had humble beginnings in the kitchen of Frank C. Mars's home in Tacoma, Washington, when he started making and selling butter-cream candy. With these early brands, goods and services for the most part were associated with specific people, a name and face, not large corporations.

In the late 1800s, as cities expanded and goods became mass-produced, trust needed to keep up with the pace and scale of industrialization. As local merchants became massive companies, person-to-person trust was no longer viable. So how were people to know the quality of the goods and services they were buying? Take beer, a product that could easily be watered down.

Established in 1777 by William Bass, the Bass Brewery grew into one of the largest beer companies in England. By the nineteenth century business was booming and in 1876 the brewery registered its distinctive ale's red triangle symbol and brand name to assure people of its quality.[23] It was the very first trademark to be registered under the United Kingdom's Trade Marks Registration Act. A new kind of branding was born.

'Brands arose as a way to compensate for the dehumanizing effects of the Industrial Age,' writes Douglas Rushkoff, author and professor of media theory at City University of New York's Queens College. 'The more people had previously needed to trust the person behind a product, the more important the brand became as a symbol of origin and authenticity.'[24]

Trust soon became centralized, top-down, opaque, controlled and institutional. Rules, regulations, auditors, market analysts, insurance and independent agencies such as the Better Business Bureau flourished, enabling people to trade beyond their immediate circle of trust. And so, by the mid twentieth century, companies faced a new challenge. With products and services now more or less standardized, how were they to stand out from the crowd?

At first, they relied on developing brand identities, recognizable by a name, logo, packaging and a tagline, which typically represented a promise – this is what this product or service will do for you. Oxo Cubes promised to 'make cooking so easy'. Lava soap promised to 'clean like no other soap'. By the 1950s, however, mass manufacturers such as Procter and Gamble, Unilever and General Foods realized these types of practical promises were not enough. The problem was, all washing powders and frozen peas do much the same job.

What would give consumers a reason to choose one product over another? Vanity, neediness, status anxiety, aspiration, nostalgia and hope, among other things, it turned out. Marketers began to tap into a whole new consumer psychology. Grandiose *brand propositions* were created, mixing functional benefits with emotional values. 'When I buy or use this brand, I am . . .' Coca-Cola wasn't manufacturing sugary drinks; its product was about making you feel 'refreshed'. Disney wasn't making movies; it was celebrating dreams. Nike, named after the winged goddess of victory, didn't sell trainers; it made you feel inspired. The idea that a brand would enable consumers to express something about themselves, something intangible, was revolutionary at the time. Crucially, and however artificially, it also bred a sense of intimacy and connection between consumer and multinational: 'Look, they care about *me*.' The result was that brand, with its fancy packaging and catchy slogans, developed enormous power and influence in our lives.

With the dawn of social media in the twenty-first century, everything changed. Marketers were hit with a seismic shift in the way trust worked with consumers. Through no-holds-barred comments and feedback, reviews and ratings, photo posts and 'likes', people started to share their experiences at scale. The person formerly known as a 'passive consumer' became a participant, a social ambassador, one who was not so easily duped and could be brutal when let down. Could Rice Krispies, the breakfast cereal, really 'help support your child's immunity'? Could a New Balance 'toning' sneaker, with hidden board technology that promised to activate the hamstrings and calves, really help burn calories as claimed?

It became much harder for brands to exaggerate or make false claims, no matter how flashy their ads. Companies had to start delivering authentic experiences and get comfortable with transparency. They had to learn how to listen, enable conversations and respond to customers' needs in real time. Brands had to let go of an era where trust could be produced and controlled centrally; by them, that is.

Fast forward to today, where conventions of how trust is built, managed, lost and repaired are once again being turned upside down. Platforms create systems that act as social facilitators. They match us with goods, rides, dates, trips, recommendations and so on. Customers have become communities, and these communities are themselves platforms that shape the ups and downs of a brand. Indeed, a recent survey conducted by Nielsen revealed that the most credible advertising comes straight from the people we know and trust. More than 80 per cent of respondents say they completely or somewhat trust the recommendations of friends and family. And two-thirds say they trust consumer opinions posted online.[25]

Compare, say, Marriott hotels to Airbnb. Once upon a time, we trusted the hotel chain; the brand was what made people feel safe to spend the night there. With Airbnb, you need to have confidence in the platform itself and in the connections between hosts and guests. In other words, trust must exist in the platform *and* between people in the community. This is one of the key dynamics that distinguishes the new era of distributed trust from the old paradigm of institutional trust.

Joe Gebbia is the thirty-six-year-old co-founder and chief product officer of Airbnb. Gebbia is a designer rather than an engineer, and he studied at the Rhode Island School of Design, where he met Airbnb co-founder Brian Chesky. I met him in 2009, when I was writing my first book, *What's Mine is Yours*. The marketplace was just starting to become more popular and Airbnb was still far from a billion-dollar idea. The founders were enthusiastic to tell the story of how they started the company from a couple of air mattresses in San Francisco; a story they have now told thousands of times.

Gebbia, Brian Chesky and fellow co-founder Nate Blecharczyk began Airbnb with the best of intentions, and had no idea of the magnitude of what they were building. Today, with the company now occupying slick 170,000 square-foot offices and with on average nearly 2 million guests staying in Airbnb rentals every night, the unintended consequences of their success have also mushroomed. One example is the concern over Airbnb's role in distorting rental prices and creating housing shortages, especially for lower-income residents. As an exercise in trust-building, however, Airbnb is a standout.

Gebbia, though passionate about design and technology and their power to bring people together, is the first to admit that Airbnb is not a technology company but is in the 'trust business'. 'We bet our whole company on the hope that, with the right design, people would be willing to overcome the stranger-danger bias,' Gebbia says in a TED talk on how Airbnb designs for trust.[26] 'What we didn't realize is just how many people were ready and waiting to put the bias aside.'

He admits there are risks. 'Obviously, there are times when things don't work out. Guests have thrown unauthorized parties and trashed homes. Hosts have left guests stranded in the rain,' he says. 'In the early days, I was customer service, and those calls came right to my cell phone. I was at the front line of trust-breaking. And there's nothing worse than those calls. It hurts to even think about them.'

Gebbia thinks of Airbnb as playing the role of the 'mutual friend' who introduces you to new friends, new places and experiences. 'We have to create the conditions for a relationship to form between two people who have never met,' he tells me.[27] 'And after the introduction has been made, we need to get out of the way.' The role Airbnb is playing may be different from Uber's, in that whom to trust is a choice that each host or guest must make for themselves, but what people want from the platform is similar. That is, we want platforms to mitigate the risk of bad things happening, and to be there for us if they do. 'The number-one thing people want from Airbnb is that if something goes wrong or not as planned, we've got their back,' Gebbia says. 'If we get that right, we are 80 per cent there.'

Alok Gupta, thirty-three, was formerly a high-frequency trader on

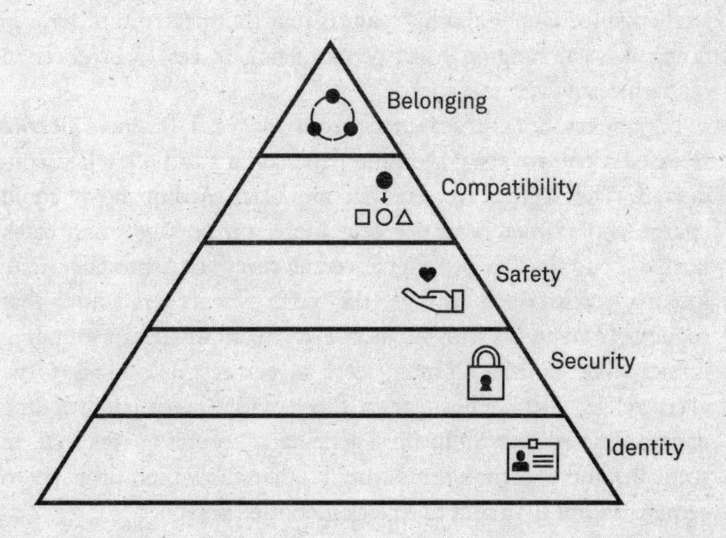

Wall Street and a research fellow in mathematics at the University of Oxford. He admits he is a big fan of observing patterns and predicting outcomes in enormous data sets. Three years ago, he joined Airbnb as a data science manager, applying his talents and thinking to a different problem: online-to-offline trust. That is, people using digital tools to meet up face-to-face. 'I think Airbnb places itself as the company which does the hard work for you, in terms of trusting the individual,' says Gupta. 'We know there's a barrier to trusting people you've never met before, but we want to fill that space, and we want to help overcome that barrier for you.'[28]

Gupta talks about the 'defensive mechanisms' Airbnb has developed that reduce uncertainty. For instance, in 2011, after the 'EJ incident' when a host infamously got her San Francisco apartment completely trashed, Airbnb introduced a 'host guarantee' covering property damage of up to $1 million per booking. In 2013, the company introduced 'Airbnb Verified ID' confirming a person's online identity by matching it to offline ID documentation such as a driving licence and a passport.[29] 'There is no place for anonymity in a trusted community,' the company wrote on its blog announcing the launch. 'Trust and verification. They just go together.' The challenge for Gupta and his team is that the spectrum of wrongdoing is vast, ranging from people using places as brothels to old-fashioned discrimination.

In January 2014, researchers from Harvard Business School released a controversial working paper on a study they had conducted. The study revealed that non-black Airbnb hosts could charge approximately 12 per cent more, on average, than black hosts – roughly $144 per night, versus $107.[30] In September 2016, looking across 6,000 listings, the same researchers found that requests from guests with distinctively African-American-sounding names (like Tanisha Jackson) were 16 per cent less likely to be accepted by Airbnb hosts than those with Caucasian-sounding names (like Allison Sullivan). Particularly troubling was that, in some instances, Airbnb users would rather allow their property to remain vacant than rent to a black-identified person.[31]

In the United States, Title II of the Civil Rights Act of 1964 explicitly prohibits racial discrimination in 'public accommodations' such as restaurants, cinemas, motels and hotels.[32] The law contains an exemption, however, for someone renting fewer than five rooms in his own home, a category that would seem to include many Airbnb hosts. 'There have been too many unacceptable instances of people being discriminated against on the Airbnb platform because of who they are or what they look like,' wrote Laura W. Murphy, a former director of the American Civil Liberties Union's Washington legislative office, who was hired by Airbnb to compile a report to serve as a blueprint for how Airbnb plans to fight discrimination on the site.[33]

In the firestorm that followed, Airbnb users started sharing stories on social media with the hashtag #AirbnbWhileBlack about their experience of bookings being denied or cancelled because of their race. For example, one user @MiQL tweeted: 'My wife & I tried to book w/@Airbnb for a vaycay. Hosts w/listed available rooms responded w/"Unavailable". White friend got "available".'[34] Personal profiles and photos, which users put together to try to project trustworthiness, have the unintended consequence of facilitating discrimination. Indeed, it seems that distributed trust is not always fairly or evenly distributed.

Ben Edelman, one of the authors of the Harvard study, says Airbnb's initial response to his findings 'was kind of denial'.[35] Nine months after the discrimination study was conducted, with pressure mounting, Airbnb released their report outlining its non-discrimination policies and promise to root out bias and bigotry. The company are trying fixes such as minimizing the prominence of user photos and trying to increase 'instant bookings' that don't require pre-approval from hosts.[36]

But why hadn't they noticed the discrimination happening on the platform? They had a blind spot. Brian, Joe and Nate are three young white American men. In other words, they hadn't personally experienced the type of discrimination many members of the Airbnb community had. 'Discrimination has no place on Airbnb. It is against what we stand for,' says Gebbia. In March 2016, Airbnb hired David King III, who had previously held a prominent diversity role at the US State Department, as the company's first director of diversity and belonging. King was given a team of talented engineers and data scientists, whose job includes identifying patterns of host behaviour and figuring out solutions to create a more inclusive platform.

But how do you go about stamping out those unconscious biases? It is difficult to remedy offline human prejudices that migrate online. You can't make somebody trust somebody. Like so much else, it's uncharted online territory. 'New systems, new structures that haven't been invented yet are needed to create environments that reduce discrimination or eliminate it,' Gebbia says.

He takes me back to the early days of the Model T Ford, which revolutionized the take-up of automobiles in the early twentieth century. 'I think the Model T has a lot of analogies to us,' says Gebbia. 'Look at early photos. It didn't have doors; it didn't have blinkers; it was missing all of these things that are needed for a safe ride, and that Ford added over the years. And sometimes I think we're like a Model T, that we haven't added our blinkers yet.'

In 1865, the British government passed a law called the 'Locomotive Act', which was a safety precaution to warn pedestrians and horse-drawn traffic of the terrifying approach of a motor vehicle. It stated that any locomotive or automobile must have a crew of three people: the driver, a stoker and a man whose job it was to walk at least fifty-five metres ahead of the vehicle, waving a red flag. The act, which later became known as the 'Red Flag Law', made it impossible for vehicles to drive more than two miles per hour in urban areas, meaning the usefulness of the new automobile was limited. The car is just one example of how, throughout history, a new technology that enables a trust leap can also introduce a new 'risky' behaviour – travelling mechanically at speed – and create a vexing challenge for lawmakers. With no precedents to go by, how do they figure out what kind of policies and restrictions will protect public interests?

*

For more than a decade, Coye Cheshire, a social psychologist and associate professor of UC Berkeley School of Information, has been studying how the internet is changing risk and trust. On a brisk autumn afternoon, I meet him in his campus office, a cosy room painted in moss green and filled with dark wood furniture and books piled high on every available surface. He makes me a cup of peppermint tea and gets straight to the heart of the matter. 'I want to help understand how humans take risks in the presence of uncertainty.'

I've long admired Cheshire's work, including a paper he wrote in 2011 called 'Online Trust, Trustworthiness, or Assurance?'[37] There, he asserted there was a difference between *interpersonal trust* (human to human) and *system trust* (human to system). Did he still think this distinction between people and technologies held true?

'Back then, systems meant things like your telephone and computer, but to be blunt, I was taking a simplistic view of technology that didn't take into account that systems are now capable of betrayal,' he says. For example, a bot in a chat room that can express feelings and moods is very different from simply cooking food in a microwave. 'Today, systems embody everything from online platforms that people are using, to autonomous agents that act on behalf of humans in ways that are blurring the line in terms of our awareness around what the machine is doing.'[38]

Cheshire admits that trust is a far more intricate business these days. 'We are working with these systems that are using complex algorithms to manage our information and make decisions on our behalf. But they are getting too complex for our brains to understand.' Consider this: to place a man on the moon in 1969 it took 145,000 lines of code.[39] Today, it takes more than 2 billion lines of code to run all of Google's internet services, dwarfing Facebook, which runs on more than 62 million lines of code.[40] 'I used to think that it was completely ridiculous to compare trust between people with trust in these online platforms,' says Cheshire. 'I'm not certain that is the case any more because in some ways we have offloaded some of our cognitive power.'

In the past, engineers would typically work on physical infrastructure projects such as roads, rail lines, gas pipelines and bridges. Today, however, they are designing new kinds of social infrastructure: online bridges that bring friends, families and strangers together. They are *trust engineers.* And one of the goals of these engineers is to get us to a place where we don't even think about the risks we are taking. It should feel like magic: you just get the right recommendation, the nearest driver, the best match, no hassle, no dramas. Yet cultivating an overload of that blind confidence can also create the opposite problem: too much trust in untrustworthy people. Think about Marc Dunton, the Uber passenger who was out with his friends at a bar when he accepted his ride with the homicidal Dalton. Dunton admits he knew there was a mad shooter on the loose and that Dalton and his car fitted the description. But surely someone on a shooting spree wasn't still on the Uber app and picking up fares? Surely the genie in the app wouldn't send me someone bad, right?

Ironically, one of the issues we face today is the speed and ease at which we are trusting. And it's not just happening on ride-sharing platforms such as Uber. Fancy a date? Download Tinder, Bumble, Happn or Tingle and zip through a fast set-up. When a profile comes up, swipe right on your phone if you like the look of them. And if they swipe right too, it's a match and you are on your way to meet up with a stranger, anywhere from one to a hundred miles away thanks to the powers of geolocation. It's accelerated trust based on a few photos and a handful of words: shopping through a catalogue of faces. It's trust on speed. And when we are in an accelerated mode of trust, we can be impulsive. It requires a conscious gear change to slow down and think twice about our decisions.

Or what about news? Have you ever shared a link without ever reading the article or watching the clip? You are not alone. As a recent study conducted by computer scientists at Columbia University and the French National Institute suggested, many people on Twitter appear to retweet news without even reading it. The researchers found that 59 per cent of links shared on Twitter have never actually been clicked.[41] 'People are more willing to share an article than read it,' says study co-author Arnaud Legout.[42] 'This is typical of modern information consumption. People form an opinion based on a summary, or a summary of summaries, without making the effort to go deeper.'

On 29 January 2017, the former White House Press Secretary Sean Spicer emphatically retweeted on his personal account a video from the satirical site *The Onion*. '@SeanSpicer's role in the Trump administration will be to provide the American public with robust and clearly articulated misinformation,' *The Onion*'s tweet joked. Around an hour later, Spicer retweeted it, adding, 'You nailed it. Period.'[43] The clip included questionable 'facts' to know about Spicer, including his former role as a senior correspondent for NPR's national desk (false), his snowy white pocket square with a quarter inch of clearance on his suit jacket, his 'defensive' speaking style and whether he has knowingly lied to the press. Could it have been that Spicer has a quirky sense of humour and was giving a sarcastic response to the video? It's possible. But it's also possible that he didn't bother to watch the video before sharing it, or to read the headline carefully enough to realize he was the punchline. In an age of 'fake news' and media propaganda, this phenomenon is more problematic and frightening than ever.

Efficiency can be the enemy of trust. Trust needs a bit of friction. It needs time. It requires investment and effort. 'Trust doesn't form at an event in a day. Even bad times shared don't form trust immediately,' says author Simon Sinek. It comes from 'slow, steady consistency and we need to create mechanisms where we allow for those little innocuous interactions to happen'.[44] Systems are becoming so seamless that we are not always fully conscious of the risks we are taking or the falsehoods we are sharing.

'I think the problem comes down to social translucence,' says Coye Cheshire. 'How much of the social interaction – our behaviours and the underlying mechanisms that enable interactions – how much of that is visible?' Not much at present, he maintains.

We need to crack open the 'black boxes' of the internet giants, to lift the veil on the behind-the-scenes operations of systems with which we interact daily and yet know very little about – and may trust too much.

Here is a simple example of how social translucence works in the physical world. If I have something valuable I want to mail, I will go to a post office rather than just drop it in a box at the end of the road and hope for the best. I will hand the parcel to the person behind the counter and pay for registered mail. A human being hands me back a paper tracking slip with a number that I can use to go online and see where my package is at any given stage of the journey. That's social translucence – there are lots of visible cues that tell me what's happening. 'With online systems, the translucence breaks down,' says Cheshire. 'Take putting your credit card information into a site. I pass along the information but it just goes in one direction and I trust the system that the details are safe.'

Online systems seem as magic as the Wizard of Oz: we do not see the army of human beings, many of them single-minded maths nerds, involved in their operation. We can't see the ghosts in the systems ranking people, places, objects and ideas, making choices and matches on our behalf. Then again, perhaps some of that ignorance is self-imposed. Many of us don't like knowing the extent to which our lives are constantly being massaged by algorithms. We prefer to trust that it's all above board.

A few years ago, a social psychologist and data scientist named Adam D. I. Kramer designed an experiment using the world's largest laboratory of human behaviour: Facebook. A team of researchers from Cornell University and Facebook, including Kramer, joined up to study 'emotional contagion' on a massive scale. Do the emotions expressed by friends via online social networks influence our moods – in other words, can emotions online be transferred to others and how?

For one week in 2012, the researchers tweaked the algorithm to manipulate the emotional content appearing in the news feeds of

689,003 randomly selected, unwitting users. Posts were identified as either 'positive' (awesome!) or 'negative' (bummer) based on the words used. In one group, Facebook reduced the positive content of news feeds, and in the other, it reduced the negative content. 'We did this research because we care about the emotional impact of Facebook and the people that use our product,' Kramer says. 'We felt that it was important to investigate the common worry that seeing friends post positive content leads to people feeling negative or left out. At the same time, we were concerned that exposure to friends' negativity might lead people to avoid visiting Facebook.'[45]

Did tinkering with the content change the emotional state of users? Yes, the authors discovered. The exposure led some users to change their own behaviours: the researchers found people who had positive words removed from their feeds made fewer positive posts and more negative ones, and vice versa. It could have been an online version of monkey see, monkey do, or simply a matter of keeping up with the Joneses. 'The results show emotional contagion,' Adam Kramer and his co-authors write in the academic paper published in the *Proceedings of the National Academy of Science* in 2014. 'These results suggest that the emotions expressed by friends, via online social networks, influence our own moods, constituting, to our knowledge, the first experimental evidence for massive-scale emotional contagion via social networks.'[46]

When the study was published, it sparked widespread public uproar. What drove the study into the spotlight weren't its findings – in fact, the effect, as the authors acknowledge, was quite minimal, as little as one-tenth of a per cent of an observed change. (Given, however, the scale of Facebook, even tiny effects can have large social consequences such as online bullying.) Instead, the outrage centred on ethics. The researchers had failed to get informed consent from the Internal Review Board (IRB) that oversees 'human subjects research', or from the thousands of Facebook users who were subjected to the manipulation. In the blaze that followed, the company argued that its 2.19 billion monthly active users give blanket consent to the company's research on the personal data it collects and stores as a condition of its terms of service. Facebook's data use policy warns users that Facebook 'may use the information we receive about you . . . for internal operations, including troubleshooting, data analysis, testing, research and service improvement'.[47] That is your price every time you log in, and the cost may be higher than you think.

Users' willingness to tick the box labelled 'Agree', on this and other platforms, has seen the blithe handover of massive amounts of once private information. 'It has enabled one of the biggest shifts in power between people and big institutions in the twenty-first century,' says Zeynep Tufekci, a sociologist at the University of North Carolina and author of *Twitter and Tear Gas: The Power and Fragility of Networked Protest*. 'These large corporations (and governments and political campaigns) now have new tools and stealth methods to quietly model our personality, our vulnerabilities, identify our networks, and effectively nudge and shape our ideas, desires and dreams.'[48]

The Facebook story is one example, and it became notorious. Shouldn't Facebook explicitly ask people to 'check the box' if they want to be made to feel happier or sadder? Comments poured in across social media. 'Does everyone who works at Facebook just have the "this is creepy as hell" part of their brain missing?' tweeted @sarahjeong a few days after the study was published. Similarly, @Tomgara tweeted: 'Impressive achievement by Facebook to snatch back the title of most dystopian nightmarish tech company.'[49] People felt they had been treated like lab rats.

What surprised many academics and researchers, including me, was the level of shock and outrage. Didn't people realize these platforms are essentially mysterious algorithms that exert immense control over what we see? Facebook is just like other content sites such as BuzzFeed and Upworthy, constantly turning one algorithmic knob and tweaking another to find an ideal ad placement, to get us to read and post more. Consider this: if you are a Facebook user, what is the statistical likelihood you have been a guinea pig in one of its experiments? According to the company, it's

100 per cent. 'At any given time, any given Facebook user will be part of ten experiments the company happens to be conducting,' says Dan Ferrell, a Facebook data scientist.[50]

Companies conduct split testing all the time, ostensibly to improve user satisfaction.* We trust algorithms to determine our Netflix and Spotify recommendations; deliver the most relevant results to our Google searches; even assess our credit score. So why all the fuss and surprise? 'The machine appears to be only a neutral go-between,' writes Cathy O'Neil in her insightful book *Weapons of Math Destruction*. In 2013, Karrie Karahalios, an associate professor of computer science at the University of Illinois, carried out a survey on Facebook's algorithm and found that 62 per cent of the people were unaware the company tinkered with the news feed.[51] So of the 2.19 billion monthly active users, 1.35 billion think the system instantly shares whatever they or their friends post.

The study struck a deep nerve – it was a reminder of how the internet churns, and where the power really lies. It illustrated the power of digital puppet masters or trust engineers constantly to manipulate our data and in different ways control our lives. And to many users it felt like they had been played; the Facebook study was considered a major betrayal of trust.

Beyond the initial brouhaha, though, the study raised a more profound question: if Facebook can manipulate a person's moods with a minor tweak of its algorithms, what else can the platform control? 'About two-thirds of American adults have a profile on Facebook. They spend thirty minutes a day on the site, only four minutes less than they dedicate to face-to-face socializing,' writes O'Neil. Nearly half of them, according to a Pew Research Center report, count on Facebook to deliver at least some of their news.[52] So this leads to the question: how else could Facebook change our minds by tweaking the algorithm? Could it change whom we vote for?

★

* 'Split testing' is the comparison of two versions of an app or web page against each other to determine which one performs better.

'Pope Francis has broken with tradition and unequivocally endorsed Donald Trump for President of the United States.' 'WikiLeaks CONFIRMS Hillary has sold weapons to ISIS.' 'Clinton runs a child-trafficking ring out of a pizzeria.' Many Facebook readers would have seen these posts and others in their news feed. A few days before the 2016 US election, I made a promise to myself not to check the site for at least a month, after a ludicrous article appeared at the top of my feed. It stated that an FBI agent suspected of involvement in leaking Hillary Clinton's emails had been found dead after apparently murdering his wife and then turning the gun on himself. Two thoughts in succession ran quickly through my mind. Was this piece of 'news' true? And then, *whom* should I go to, what news source, to find out the real truth? All these stories were of course hoaxes, typifying the fake news and conspiracy theories that plagued the 2016 election.

A recent study by BuzzFeed found that 38 per cent of all posts from three of the largest hyper-partisan right-wing Facebook pages, such as Eagle Rising, contained a mixture of true and false, or mostly false information, compared to 19 per cent of posts from three hyper-partisan left-wing pages, such as Occupy Democrats. Cumulatively, the audiences of these pages are in the tens of millions. The top five fake news items in the last weeks of the election were all negatives for the Clinton campaign – in other words, the Facebook algorithm picked a side – it's not neutral. And the spread of fake news is far more common on the right than it is on the left. 'These findings suggest a troubling conclusion: The best way to attract and grow an audience for political content on the world's biggest social network is to eschew factual reporting and instead play to partisan biases using false or misleading information that simply tells people what they want to hear,' writes Buzzfeed.[53]

It's an ironic turn of events, given that in May 2016, Facebook came under fire from Republicans and critics for allegedly suppressing conservative-leaning stories in its trending news section, curated by a small team of human editors. Facebook ended up replacing those human editors, accused of party bias, with software, but the plan clearly failed. And, tellingly, the top twenty fabricated election stories on Facebook netted more engagement than factual stories from mainstream news sources.[54] In September 2017, the US Senate intelligence committee launched an official investigation into how Russia used social media to influence the result of the presidential election. Representatives from Facebook, Google and Twitter were obliged to submit evidence about relevant activity on their platforms. Twitter provided a list, sixty-five pages long, with the handles of some 36,746 Russian-linked bots that tweeted a total of 1.4 million times. The company estimates these tweets were viewed 288 million times.[55]

Facebook also admitted to lawmakers that between June 2015 and August 2017, 11.4 million Americans definitely saw Russian IRA (its troll farm) advertisements, which ranged from 'like and share if you want Burqa banned in America' to claims that 'Hillary is Satan, and her crimes and lies have proved just how evil she is. And even though Donald Trump is not a saint by any means, at least he is an honest man, and he cares deeply for his country.'[56] At the bottom of the ad was a press 'like' to help Jesus (Trump) win. The most successful ads were clicked on and shared by almost a quarter of the people who saw them. This means, according to Facebook, that up to 126 million residents (or almost half of the US population) were likely to have seen a Russian-linked post.[57]

Were we naive? As unprecedented numbers of people channelled their political energies and beliefs into social media, shouldn't we have foreseen the way the platforms could become vulnerable to manipulation and the spread of misinformation? Probably, but most of us failed to imagine the imaginable.

In a recent interview on the Netflix show *My Next Guest Needs No Introduction With David Letterman*, Obama reflected on his first year out of office. He never mentioned Trump by name, but he did candidly discuss the 'pervasive divisiveness' in society and how it is exacerbated by social media.

'One of the biggest challenges we have to our democracy is the degree to which we don't share a common baseline of facts,' Obama said. 'What the Russians exploited – but it was already here – is we are operating in completely different information universes. If you watch Fox News, you are living on a different planet than you are if you listen to NPR.' Put another way, we don't have a shared sense of reality and that can be seriously played on.

In the wake of the President Trump's unexpected victory, Mark Zuckerberg initially denied the allegations that foreign trolls and politically motivated bots sowing discord on Facebook could have influenced the election results. 'Personally, I think the idea that fake news on Facebook – it's a very small amount of the content – to think it influenced the election in any way is a pretty crazy idea,' he said a few days after the election.[58] A few months later, however, he had significantly shifted his stance from the initial *Who, us?* shrug reaction. In February 2017, he published a 5,700-word manifesto on his Facebook page.[59] It sounded a bit like a grandiose State of the Union address for 'bringing us closer together', outlining the immense challenges facing the world today, from terrorism to climate change, pandemics to online safety. 'Every year, the world got more connected and this was seen as a positive trend. Yet now, across the world there are people left behind by globalization, and movements for withdrawing from global connection,' he wrote. 'There are questions about whether we can make a global community that works for everyone, and whether the path ahead is to connect more or reverse course.'

Zuckerberg dedicated approximately 1,000 words to how Facebook has become a hotbed for fake content and why this is leading to increased polarization. 'If this continues and we lose common understanding, then even if we eliminated all misinformation, people would just emphasize different sets of facts to fit their polarized opinions. That's why I'm so worried about sensationalism in media.' The CEO then went on to describe a plan, albeit a vague one, for how Facebook proposes to deal with issues on the platform. But can the internet giant really wage war on disinformation and quash bogus memes? 'Don't believe everything you read in the newspapers,' the old saying used to go. Now it is 'don't believe everything you read on Twitter or Facebook'. According to the latest Edelman report published on 22 January 2018, 63 per cent of the 33,000 respondents said they no longer knew how to tell good journalism from rumour or falsehoods. In the UK, around 70 per cent of Britons believe social media companies are not doing enough to stop extremist content being shared or to tackle illegal behaviours on their platforms. 'Trust is only going to be regained when the truth moves back to centre stage,' commented Richard Edelman in the report's summary.[60]

Moving forward, Facebook will check whether people are reading the articles before sharing. If they are, those stories will get more prominence in the news feed. Users can flag posts they think are fake or suspect, helping Facebook detect the most blatant posts. Artificial intelligence and algorithm analysis will also be used to flag content and detect dangerous falsehoods. But how do you decide what is intentional disinformation and what is, say, a bit of an exaggeration? Who decides what the truth is? If it's Facebook doing the deciding, we are awarding them even more power to set the agenda. I'm a huge advocate of free speech, open democracy and online dialogue that mirrors the exchange of opinions that happens offline – at work, at home and even in classrooms. The issue is that it has become a free-for-all, a corruptible beast that we can't, or haven't, yet learned to control. We don't have the tools to deal with the scale of fraught challenges created by a new world of self-reference and 'hands-off' proprietors.

'Accuracy of information is very important. We know there is misinformation and even outright hoax content on Facebook,' Zuckerberg said in the February 2017 letter. 'In a free society, it's important that people have the power to share their opinion, even if others think they're wrong.' He added, towards the end: 'Our approach will focus less on banning misinformation, and more on surfacing additional perspectives and information, including that fact checkers dispute an item's accuracy.'

To be clear, I do not think that the data scientists, researchers and engineers at Facebook are bad people intentionally gaming the political system. But while Facebook isn't responsible for untrue stories *per se*, the company wrote the code that meant the bogus news items appeared more prominently over the picture of a friend's child, a funny video or a genuine news story. Is it up to Facebook to emphasize and provide information about the original news sources for news articles? Should they do more to contain hoax stories? Is it the fault of a Facebook business model that depends on clickbait? I think yes, to all of the above. But perhaps the betrayal lies in the system itself, a system that makes it easy for almost a third of the world's population to gossip and gripe, share and like, even if the content is false, and without proper checks and balances or any real redress.

The online landscape is vastly populated and yet, all too often, empty of anyone to take charge or turn to when it counts. It's rather like when you're a teenager and you throw a party while your parents are away. At first, the freedom is thrilling but there comes a point, somewhere after the tequila slammers, window smashes and the gatecrashers, when you start wishing there was a responsible adult in the house.

Facebook insists it is not a media company but merely a 'neutral technology pathway' facilitating connections between people. It is a misconceived and dangerous position. It is a media company with enormous influence in shaping someone's worldview about whom to trust. The platform controls how a quarter of the world's population shares and receives information. No single entity should have that kind of domination. And it is profit-driven. In May 2017, Facebook reported that 98 per cent of its quarterly revenue came from advertising, up from 85 per cent in 2012. In other words, it's in the company's interests to keep our eyes glued to the screen, no matter what the content. Zuckerberg argues the ad-based model is fundamentally serving the ethical mission of Facebook – to provide free services to as many people as possible. The complication is that you, the user, are also the product, which results in a massive conflict of interest for Facebook. When there's a ton of money to be made in monetizing your data, how much trust can you place in

Facebook's expressed good intentions to bring the 'world closer together'?

The questions around Facebook may sound different from the ones raised in the days after the Kalamazoo killings but in fact they are remarkably similar. They are symptoms of a systemic problem: a crisis of accountability. That's not a technological problem, it's a cultural one. When it comes to trust in distributed systems, we need to know who will tell the truth about a product, service or piece of news, and who to blame if that trust is broken.

Where does the buck stop? In this new era, people are still working that out.

With traditional institutions, the picture was clearer. For instance, when, say, your Barclays bank account was hacked, the bank would reimburse you. But when an online cryptocurrency fund such as the Decentralized Autonomous Organization (DAO) runs into trouble, there is no central ombudsman or traditional institution to turn to. (Instead, pandemonium rules, as we'll see in chapter 10.) We are in unmapped territory, scrambling and fumbling around for mechanisms that can replace institutional trust, and at the same time looking for ways to improve the old world's own shortcomings in matters of accountability. Platforms, meanwhile, are trying to figure out their role in it all – mere facilitators in bringing people together, or something more? And who's responsible when things go wrong?

Going back to the analogy of the early days of the car, it took decades to create norms like traffic lights, stop signs and even something as simple as the concept of road lanes. 'We will look back one day and laugh, "Imagine a car without blinkers",' Gebbia says. 'Then we will realize how far we have come in this new era of trust.'[61] Even at that point, with some clearer guidelines governing our coexistence in a world of distributed trust, we will still occasionally crash into each other. No system is foolproof. The hope is that it will be a minor collision, not a fatal accident.

5
But She Looked the Part

My first lesson in the dangers of trusting strangers came in the spring of 1983, not long after I turned five, when an unfamiliar woman entered our house. She was neither a family member nor friend. Her name was Doris, she was in her late twenties and she was starting as our mother's help. She came from Glasgow and had a thick Scottish accent. Her o's sounded like 'ae' and she rolled her r's. It was gently lilting, almost sing-song.

Doris had a mop of mousy brown hair and wore thin steel-rim glasses. She was plump with a ruddy face. She was the type of person you could imagine going for a brisk walk on a cold day and then sinking into a comfy chair, content with a cup of tea and a shortbread biscuit.

She arrived at our house wearing her 'Salvos' uniform. It was a navy suit with big silver S's embroidered on the collars, complete with a bonnet-style hat. Doris said she belonged to the Salvation Army because she enjoyed helping people. She didn't bring many belongings with her, although I remember the tambourine she kept at the side of her bed.

My mum had found Doris through a posh magazine called *The Lady*. A young aristocrat named Thomas Gibson Bowles, who also started *Vanity Fair*, founded the magazine in 1885. If you watch *Downton Abbey*, you will have heard of *The Lady*. It is the place where high society – including the Royal Family – seeks domestic staff, from gardeners to butlers to nannies. You won't find any celebrity tittle-tattle or sex stories in the magazine whose tagline is 'for elegant women with elegant minds'. The lead articles from a past edition included 'Capture the Style that Wooed a King', followed by 'Where to Find Bluebells in Bloom'. There was even a recipe to make teatime Bakewell tarts. You get the picture.

My family is not high society, far from it. So I was intrigued as to why on earth my mum had advertised for help in *The Lady*. 'I was starting my own business and feeling nervous about hiring someone to look after you,' she explains, all these years later. 'I remember thinking if the Royal Family uses *The Lady* to find help, it must be reliable and the best.'

Doris replied to Mum's advertisement. In those days, you would send a formal letter expressing interest in the position and a photograph of yourself. An interview would follow. As Doris lived in Scotland, Mum interviewed her over the phone. 'I remember her strong Scottish accent,' Mum tells me. 'She said all the right things. She told me she was a member of the Salvation Army and had worked with kids of a similar age. But, honestly, she had me at "hello".' After their chat, Mum called the references Doris had given and was satisfied they were all impeccable.

Doris lived with us for just over ten months. She was for the most part a good nanny – cheerful, reliable and helpful. There was nothing strikingly suspicious about her, except for one thing. After school every Wednesday, she would drive us to a block of council flats in Edmonton. The building was one of those dark grey concrete highrises. It sat close to the North Circular, a busy ring road in London. An odd man in his fifties, balding, lived in the flat. And so did a young baby. The flat was dingy and things were always strewn everywhere. I still remember the dowdy wallpaper and damp, musty smell. Doris would spend the entire visit cuddling the baby.

I told my parents I didn't like going to this strange flat, to see this strange man. Doris insisted that she was visiting the only family she had in London. Her 'uncle' made us nice tea and we liked playing with the baby. The weekly after-school trips continued.

On one of these visits, I noticed there were lots of bottles of expensive-looking perfume on the table; they looked just like the ones Mum had in her own bathroom. I mentioned it to my parents. Funnily enough, it was one of the first times I remember my parents not believing me. I was a dreamer. I had imaginary friends and made up elaborate plays. They told me to stop making up stories about Doris. That it wasn't nice. So nobody suspected anything was amiss. Or not until Doris's Uncle Charlie died, supposedly.

One night, around nine months into her stay, Doris didn't come home. When she did return she explained her Uncle Charlie had suddenly died of a heart attack and she had rushed back to Edinburgh for the funeral. Doris's mum happened to call our house later that afternoon. My parents naturally offered their condolences. 'Her mum had no idea why,' Dad tells me now. 'Doris's mother said the brother was alive and kicking. In fact, he was sitting in the armchair having tea right next to her in her lounge.'

Dad confronted Doris. She said her mother was in shock and must have forgotten. 'I told her it was highly unlikely you would forget your brother dying,' Dad recalls. Doris finally confessed that she had lied because she had really gone to the VJ Day veterans day parade to see Princess Diana. My parents thought it was slightly odd but Doris was obsessed with the Royal Family so it was plausible. She continued living with us.

The series of events that subsequently unravelled sounds totally unbelievable. You'll have to take my word that it's true.

We had lovely neighbours at the time called the Luxemburgs. They had kids of a similar age and also an au pair. Doris spent *a lot* of time with her. Around a month after the Uncle Charlie incident, Mr Luxemburg knocked on our door late one evening. He told my parents that he had just thrown their au pair out. 'Philip said he found out that she had been involved in running some kind of drugs ring in North London with Doris,' Dad relates. 'They had even been in an armed robbery and he believed Doris was the getaway driver.' The car, it later turned out, was our family's silver Volvo Estate.

At this point, my parents decided to search Doris's room. They found plastic bags full of credit-card statements and thousands of pounds' worth of unpaid bills. In a shoebox under her bed, she had stuffed piles of foreign currency, stolen from my parents' home office. Now on high alert, my dad stood on guard by our front door all night with a baseball bat. He was frightened Doris would come home. Thankfully, she didn't.

The next morning, Dad went to the police first thing. He drove with them to the flat we had been visiting on Wednesdays, visits my parents knew about, even if they had been misled about the true circumstances. 'There was a big hole in the front door that somebody had tried to kick in,' Dad recalls. The weird guy was there, the supposed 'uncle' who gave us tea. (Turned out the 'uncle' was Doris's boyfriend and the baby was their child.) He had a big iron bar on the table. Doris never returned to our house.

'Even as I retell this story I feel sick to my stomach,' Mum says now. 'I left you in the care of a serious criminal. And it took us so long to know who she really was.' My parents never hired anyone through *The Lady* again. Instead, they asked their friends for referrals.

Looking back, what would they have done differently? 'I wish we had asked Doris more and better questions,' Mum says. 'I wish we had known more about her.' She now realizes the impeccable referees could just as well have been Doris's friends, family or even 'colleagues' in her drugs ring. And the Salvation Army was a total cover story.

My parents thought they had enough information to make a good decision about Doris, even though in retrospect there was *a lot* they didn't know about her. There was a *trust gap.* And that raises an essential point when it comes to trust: the illusion of information can be more dangerous than ignorance. As the Italian social scientist Diego Gambetta beautifully put it, 'Trust has two enemies, not just one: bad character and poor information.'[1]

It would be helpful if the likes of Doris wore labels saying, 'Be warned, I am a con woman and serial liar.' But they don't, and of course it's in the nature of such a person to be convincing. My parents clearly made a very, very poor decision. Yet they are generally smart, rational people with good judgement. What went wrong?

*

Baroness Onora O'Neill is a philosopher, a professor at the University of Cambridge and a cross-bench member of the House of Lords. Now in her late seventies, she has written extensively about trust and, crucially, how trust is misplaced. She explores that theme in a TED talk, while also challenging the conventional, simplistic belief that as a society we have lost trust and ought to set about rebuilding it.[2] More has to be better, right?

'Frankly, I think rebuilding trust is a stupid aim. [Instead] I would aim to have more trust in the trustworthy but not in the untrustworthy. In fact, I aim positively to try *not* to trust the untrustworthy,'[3] Baroness O'Neill tells her audience, with understated dry humour.

Her point, however, is deadly serious. Trust is not the same as trustworthiness.[4] Encouraging generalized trust simply for the sake of creating a more 'trusting society' is not only meaningless, it's dangerous. For one thing, people are already inclined to want to trust blindly, particularly when greed enters the picture. The Bernie Madoff scandal is a classic case. Think of all the tens of thousands of investors who placed their savings with the aptly named Madoff, who made off with their money in an elaborate $65 billion Ponzi scheme that ran over decades.[5] Why did investors trust him about something too good to be true? Mostly because Madoff was charming and moved in the same country club and Jewish social circles as they did, in Long Island and Palm Beach. He was a long con, a person who had built up his reputation over years. Indeed, he was known for being a generous, charitable man (it just turned out to be with other people's money). And besides, his own family, close friends and showbiz names such as Steven Spielberg and Fred Wilpon, owner of the New York Mets, had invested with him. The guy had to be sound, didn't he? No, as it turned out.

As O'Neill notes, Madoff is an example of too much trust in the wrong place. Instead, all of us making decisions about trust should be looking at the who, where and why of *trustworthiness*. Who deserves our trust, and in what respects do we need them to be trustworthy? For instance, if I asked, 'Do you trust your dentist?'

that in itself is not a helpful question. You might sensibly respond, 'To do what?' 'Intelligently placed and intelligently refused trust is the proper aim [in this life],' the baroness reiterates. 'What matters in the first place is not trust but trustworthiness – judging how trustworthy people are in particular respects,' says O'Neill.[6]

How well do we carry out that logical goal in practice? It's not always easy.

My parents' decision to trust Doris came down largely to their personal judgement and blind faith. They wanted, even needed, to believe that what she was saying was true. Their judgement of Doris was also influenced by *trust signals*. These are clues or symbols that we knowingly or unknowingly use to decide whether another person is trustworthy or not. The Salvation Army, Scottish accents, *The Lady* magazine, Doris's cheery appearance, her references and even her steel-rimmed glasses were all trust signals my parents used to make a decision. Trust signals supposedly give us the ability to 'read' each other. They give us reasons to trust someone or ways to demonstrate our own trustworthiness. But it's still a bet, of sorts. 'Like all gambles, assessing trustworthiness is an imperfect endeavour; there's always a chance you're going to come up short,' writes David DeSteno in *The Truth about Trust*.[7]

Some signals we literally 'give off', such as our clothes, our face and our accent. Indeed, studies have shown that the Scottish accent is perceived to be the most trustworthy in the United Kingdom ('Scouse' is perceived to be the least).[8] Other trust signals are non-verbal but still visible, including our posture or gestures such as a nod, smile, twitch or an averted gaze. Despite the admonition not to judge a book by its cover, these first impressions are insanely influential when it comes to trusting someone.

Jon Freeman is an assistant professor of psychology at New York University and director of its Social Cognitive and Neural Sciences Lab. He studies what he calls 'split-second social perception'. When you see someone's face, you make snap judgements, within a tenth of a second, about their traits, including how trustworthy they are.[9]

Freeman wants to understand why our brains take these kinds of mental shortcuts.

Freeman, in his late twenties, is a rising academic star. On a typical day, you will find him wearing slim-fitting slacks, a navy blue button-down shirt and nerdy-cute tortoiseshell glasses. Looking at him, you might instantly label him an 'academic'. The scientist behind 'blink' stereotyping likes it that way. And, in this case, you'd be right.

A few years ago, Freeman and his colleagues devised an experiment to see if there was such a thing as appearing trustworthy. Participants were shown pictures of distinctly different male faces of different ethnicities. They were asked to rate how trustworthy or untrustworthy they thought the people in the pictures were. The results were clear – your brain thinks it knows a trustworthy face.[10] Humans are inherently wired this way. When our ancestors were approached by a stranger, they needed a rapid-response mechanism. Friend or foe? But the same rapid response in our day-to-day reactions can lead us to make biased trust decisions based on stereotypes.

The researchers devised a second experiment in which they digitally altered pictures of the same person, evolving the images gradually from looking *slightly happy* to *slightly angry*. The study found that people with upturned eyebrows, pronounced cheekbones, big baby-like eyes and an upward curving mouth – even if they weren't overtly smiling – are more likely to be perceived as trustworthy. Those featuring sunken cheeks, a downturned mouth and eyebrows – even if they weren't obviously frowning – are more likely to be perceived as untrustworthy. So the trick to appearing trustworthy? Look *slightly* happy. Just like Doris did. The problem is, however, there is no evidence that people with those 'trustworthy' features are in fact more trustworthy.

You can't consciously control whether you perceive a person to be trustworthy or not; your brain does it for you automatically, but not always accurately. And our early assumptions about people can be difficult to budge. That's astonishing and a little frightening.

Trust can therefore be easily misplaced. Indeed, the art of the con man is to give off the right signals and appear trustworthy.

Of course, trust signals are not just based on looks and first impressions. Clear symbols of status or authority, from a white lab coat to a police officer's badge, are also trust signals, for some. Uniforms can be powerful shortcuts for enabling trust. For instance, if my doorbell rings and I look through the peephole and see a stranger in a postal uniform, I will open the door. The postal uniform is a recognizable symbol that reassures me in my decision. Well-known brands rely on the same dynamic. When I am overseas in a country with dodgy drinking water, I'll buy bottles of Evian or some other brand I know. Why? The name and packaging make me believe the water inside is safe. I don't have to trust the person selling me the water, I trust the brand.

Trust signals can also come from the endorsement of third parties. Doris used her association with the Salvation Army, one of the most trusted charities in the UK, to appear trustworthy. She was deliberately sending false signals to dupe my parents. We may not outright lie like Doris did, but think of the number of times we drop in an association with a trusted brand or institution to establish our credibility. I do it all the time when I meet new people: 'Oh, I teach at Oxford University' or 'You might have seen my work in *The Economist*.' I am not bragging (okay, maybe a bit, sometimes) but intentionally drawing on these signals to build trust. Going back to the definition of trust, *The Economist* and Oxford reduce uncertainty about me for other people. They build confidence in the unknown.

We use institutional trust signals to help us make all kinds of decisions in our lives. For instance, imagine you have to choose between two lawyers to handle a matter. One has a law degree from Whittier Law School in California (ranked as one of the worst law schools in the United States) and the other has a degree from Harvard Law School. Removing price from the equation, most people would choose the Harvard-trained lawyer. Similarly, 'You can't get fired for hiring McKinsey' is an adage that has been muttered among management for years as a reason for hiring the world's most prestigious consulting firm. In both these examples, we are trusting the elite reputation of the institution, not the trustworthiness of the individual.

Trust signals are changing in the age of distributed trust. Consider occupational licences: in the 1950s, fewer than 5 per cent of American workers needed a licence to do their job. Today, more than 1,000 professions – approximately a third of all occupations – in the United States require a licence. In some states you need a licence to be a tree surgeon, fortune teller, florist, horse masseuse, make-up artist, ferret breeder, falconer and even a hair braider.[11] Of course, occupational licences are necessary to enforce standards for high-risk professions that require you to put your life in someone else's hands. I like knowing that the pilot flying the plane or the doctor at the hospital are regulated and licensed. But professional licensing rules have become excessive and don't necessarily help us to make good decisions as to whom we can trust. Do I really need my hairdresser to pay $2,500 per year and to have no fewer than five sinks and at least ten workstations before I can trust them to cut my hair? Surely ratings and reviews left by previous customers would be a better, or at least an equal, indicator of their talent?

In some instances, new trust signals will be used alongside institutional ones. Think about lawyers. Law degrees or bar memberships are certainly important trust mechanisms, but here's the thing: once earned, these qualifications do not change based on a lawyer's *actual* performance (unless, of course, he or she is removed from the bar for unethical behaviour). Who judges the judgement and quality of skills of professionals? Indeed, I once made a very bad decision in hiring a lawyer who had graduated from an Ivy League law school and belonged to a top-tier law firm. A friend had recommended him to handle a matter. He turned out to be unresponsive and, frankly, incompetent. If I was hiring a lawyer today, I might go to UpCounsel, an online marketplace that matches clients with high-quality attorneys, based on particular skills, pricing and availability. There I could find a suitable one at the push of a button, just as easily as I could hire an Uber ride.

On every lawyer's profile, it lists qualifications, years of experience and areas of specialization. Clients can also see in real time how much their attorney is billing and what for. Plus, after the transaction, lawyers are rated by clients with one to five stars, just like on eBay, and given detailed reviews. 'Reviews are one of the most important things in my online profile,' says Seth Weiner, a popular attorney on UpCounsel, who graduated from Columbia Law School and left life in the big city firms to work as a solo practitioner.[12] 'If 250 people have been happy with me, it stands to reason that the next person will be as well.' UpCounsel is using online reputation as a means of solving problems of trust. The fascinating aspect of these real-time ratings is how they help hold people accountable for performance once they are in the profession. They allow us to see beyond the more obvious hallmarks of 'goodness' (such as Harvard degrees and fancy offices) that could be totally at odds with the quality of work they *actually* perform.

So technology might hold people more accountable, but to what degree can technology enhance our assessment of who is trustworthy and who isn't – whether choosing a nanny, hiring a lawyer or ordering a taxi? Would my parents have made the same mistake in today's digitally connected world? Maybe not. Technology today can dig deeper into who we are than ever before. Can an algorithm determine who can't be trusted, better than us? I knew the right person to ask.

Lynn Perkins is the forty-three-year-old co-founder of UrbanSitter, an online marketplace that connects families with babysitters on the internet. She lives in San Francisco and she talks fast, really fast. 'I love to help people find the perfect restaurant, the right vacation for a friend's honeymoon or a new job,' she says. 'I've matched friends needing apartments with roommates and three couples who are now married.' In short, Perkins is a connector.

In 2008, after the arrival of twin boys, Perkins decided to take time off from her high-flying career in investment banking and real-estate development. Long hours in the office were replaced with lots of time with other mums. The conversations were somewhat predictable. How do you get your child to sleep more? Why are they so picky about what they eat? Why don't they listen to me? And so on. Perkins also noticed, however, the inordinate amount of time mums spent venting about the shortage of reliable childcare. 'They'd say they would rather skip going out on a Friday night with their husband if they couldn't find a sitter their friend had already used,' Perkins tells me. 'It was incredibly revealing about what they were looking for and currently couldn't find.'

Perkins herself had experienced the nuisance of one too many sitters cancelling at the last minute. 'It was around the time when companies like OpenTable, for restaurant reservations, and Airbnb were starting to crop up,' she says. 'But there wasn't any kind of on-demand marketplace for babysitters and nannies.' Why, she wondered, could you book a table in thirty seconds but not a sitter? She suspected it was down to trust.

In the early days of UrbanSitter, Perkins did something simple but smart – she borrowed trust from organizations parents already trusted. She went to local music classes, clubs such as Big City Moms, Little Stars Soccer teams, elementary school groups, you name it, and found all the babysitters the parents in the groups were currently using. Perkins convinced the best sitters she met through these organizations to put their profiles on UrbanSitter. But it wasn't enough. When Perkins first started UrbanSitter in 2011, her closest friends and the investors she pitched to were convinced it would never work. 'Are you crazy?' was a common response.

'They just could never imagine using a service where people would find their care provider online,' says Perkins. The doubters wondered how finding a sitter online could be better and safer than asking for personal recommendations from friends. 'One thing everyone wanted to know was whether my team had met and interviewed every single one of the individual sitters. It was something they thought you just had to do to build trust.'

Even then, people simply couldn't imagine that social networks could not only deliver all that essential information but also make a better job of it. As is now the case with many online services, you have to join UrbanSitter via Facebook or LinkedIn. If a person has fewer than five 'friends', it's a red flag that it's a fraudulent account. But the real power of the Facebook login is that it can unlock the value of established personal connections. It reveals how we are connected to others. Whom do you know that I know, be it direct Facebook friends, friends of friends, people we went to school with or who worked at the same company. Trust established in one group and context can travel and spread to another.

On UrbanSitter, when you go to book, you can see how many 'friends' have previously booked or are in some way connected to that sitter. These connections make us feel more comfortable and confident about our decisions. They reduce the unknown. The collective wisdom of the crowd is enhanced by the wisdom of 'friends'. It's social proof on steroids.

The late John Keith Murnighan, when he was a professor in social sciences at the Kellogg School of Management, set out to explain what causes us to trust people we do not really know. Specifically, he was interested in the role that 'friends' play in stimulating feelings of trust for a stranger.[13] He conducted a series of experiments based on the famous 'trust game' that was originally designed by behavioural economists in 1995. In the game, there are two participants, a sender and a receiver, who are anonymous strangers. Both are given a certain amount of money, let's say $100. The first player can send any portion of the $100, or none of it, to the second player whom they will never meet. The first player is told that whatever amount they send will be tripled, by the experimenter, for the receiver. The receiver then has to make a similar decision: how much of the tripled amount should they return to the sender? Player A can therefore either potentially turn a profit or lose everything. The point of the game is that sending a large amount of money indicates a high degree of trust.

Before volunteers participated in the trust game, Murnighan and his team asked them to provide the names of people they trusted and people they distrusted and the reasons for their feelings. The researchers quickly flashed these names for mere milliseconds to subliminally prime the study participants. It was too fast for anyone to be able to recognize the names that had been flashed. After that, the participants played the classic trust game. The results were stunningly clear.

Participants who had subliminally seen the names of people they trusted, sent on average nearly 50 per cent more to the anonymous receiver than the participants who saw the names of people they did not trust. 'We found we could stimulate feelings of trust for a stranger without people even realizing,' wrote Murnighan, an outcome he found 'both exciting and scary'. 'Imagine a fanatic fan of Elvis Presley. If I know someone is a huge fan of Elvis, I might casually drop Elvis's name to activate more trust in me. There is clearly a risk of manipulation.'[14]

The results of the game help explain how the likes of Madoff managed to deceive so many people. His client list included the rich and famous and his own friends and family – his brother, his sons and their wives. When new investors spotted close relatives on the investment list, it was a powerful trust signal that Madoff himself was trustworthy. Whether Madoff intentionally or inadvertently used the friends and family connections does not matter. The point is, these relational cues can provoke automatic trust that can be very dangerous, especially in cases such as Madoff's where we might lack the time or expertise to make a deliberate and proper evaluation.

But Murnighan's experiment also sheds light on the power of online social connections – the wisdom of 'friends' can automatically enhance our ability to trust people we do not know.

Interestingly, Lynn Perkins wrongly assumed the most influential social connection would be between parents. Instead, it was between sitters. 'Over time, parents value the connections UrbanSitter can surface between sitters; they want to book the friends of the sitter they really liked,' she explains. When you think about it, this is how trusted referrals typically work. Say an entrepreneur asks me for the name of the designer I trust. I introduce them to Amy Globus, whom I have worked with for years, but she is too busy to take on their work. The entrepreneur is then likely to ask Amy to recommend another designer rather than coming back to me for another suggestion. In other words, trust really lies within the group with the expertise (the babysitters) rather than the group with a similar need (the parents).

The social graph* is information manna when it comes to trust. We are now using digital tools to approximate old-fashioned ways of finding people we could trust, through referrals and close connections, but in ways and on a scale never possible before. 'It's a digital recreation of the neighbourly interactions that predefined industrial society,' writes Jason Tanz in an excellent piece on digital trust in *WIRED*. 'Except now our neighbour is anybody with a Facebook account.'[15]

The strange and beautiful truth about the social graph is that it shortens the distance between any two people in the world. Think of it as the string tying together the arbitrary connections between humanity. In 1929, the Hungarian author Frigyes Karinthy first wrote about the theory of the 'Six Degrees of Separation', later popularized in a play by John Guare, claiming that everybody on this planet is separated by only six other people. This is where the 'Kevin Bacon game' comes from – it would take the average person six or fewer acquaintances to be connected to the actor. Today, that number has shrunk significantly, at least for the 2.19 billion active users on Facebook. In 2016, Facebook crunched their social graph and determined that the degrees of separation number is 3.57.[16] In other words, each person on Facebook is connected to every other person by an average of three and a half other people. It means that, despite Facebook's vastness, it feels intimate. It gives us the sense of

* The 'social graph' refers to the connections between you and the other people, places and things you interact with in an online social network.

connection and trust we used to get from real-life communities and neighbourhoods. Its size, however, is critical to its usefulness.

When my parents decided to trust Doris, they based their decision on faith – they believed what she and her references claimed. In the past, we had to make a lot of decisions based on blind faith or personal experience, but today we can base them on *collective experience* – the experiences other people have shared through reviews and social networks.

I don't need to have personal experience with a nanny I hire through UrbanSitter to assess if she or he is trustworthy; I can benefit from other people's experiences. It's a dynamic often referred to as *indirect reciprocity* – and it can speed up the process of trust. Today, on UrbanSitter, the average sitter responds in less than three minutes, down from twenty-three hours five years ago. The average time a parent takes from either posting a job or doing a search for a sitter to accepting the person they trust to look after their kids is less than ten minutes, down from twenty-three hours when UrbanSitter first started. It shows that, online, questions of trust can be settled very fast. And this process is only going to get faster, from minutes to seconds.[17]

There is just the small matter of things going wrong now and again in this speedy and efficient brave new world.

It is almost impossible to get entrepreneurs, people like Lynn Perkins, to tell you the precise number of bad incidents, minor or serious, that happen on their platforms. I have prodded and cajoled to try to get actual data but instead you tend to get a generic response. 'They do happen but they are extremely rare,' insists Perkins. Take Wendy, a mum living in Seattle who hired a sitter to look after her six-month-old daughter one night in May 2016. The day after, she received a call from her bank about a cheque with a dodgy signature. It turned out the sitter had gone through Wendy's drawers and stolen a cheque that she had written out for $1,300. 'As soon as the parent let us know what happened, the sitter was blocked from our system,' Perkins tells me. Bad apples will make the cut, it's inevitable, but it's also likely they'll be exposed more quickly.

'She seemed so trustworthy' is something we commonly say after trust has been misplaced. On the flip side, you have probably been on the receiving end of someone saying 'I trust you'. But what do these statements mean? And, coming back to Baroness O'Neill's point, what is it that we should be basing our trust upon?

Turns out, there's a relatively simple formula to trustworthiness that goes beyond 'but he had kind eyes' or 'she looked the part'. It doesn't matter if you are deciding whether to trust an estate agent, a lawyer or a babysitter, the four traits of trustworthiness are the same: Is this person competent? Is this person reliable? Does this person have integrity? Is this person benevolent?[18]

Competence comes down to how capable a person is to do something. Does he or she have the skills, knowledge and experience to do a particular role or task, be it cut my hair, mind my children or fly me to Uzbekistan.

Reliability comes down to a person's consistency in doing what

TRAITS OF TRUSTWORTHINESS

ABILITY

Competence

Reliability

CHARACTER

Integrity

Benevolence

they said they would do for you. Ultimately, it's about you knowing, 'Can I depend on this person? Will he or she follow through?'

Integrity is about honesty and fairness. 'What are their interests and motives towards me? What do they gain by lying or by telling the truth?' In the words of Dr Seuss, people with integrity 'say what they mean, and mean what they say'.[19]

Benevolence comes down to empathy and goodwill. How much does this person care about the goal or commitment?

Political scientist Russell Hardin eloquently argues that trust is really about *encapsulated interest*, a kind of closed loop of each party's self-interests. He argues that if I trust you, it's because I believe that you are going to take my interests seriously – whether it be for friendship, love, money or reputation. Why? You won't take advantage of me because it benefits you not to do so. 'You value the continuation of our relationship, and you therefore have your own interests in taking my interest into account,' Hardin writes in *Trust and Trustworthiness*.[20] For instance, I trusted the estate agent who recently sold my house to get a good price not because she was nice or cared about me but because her commission was directly tied to the sale price. That's encapsulated interest. Or as economist Adam Smith would put it, the estate agent's future payoff is a strong enough incentive for her and good enough reason for me to trust her.

We often mentally ask of someone, 'Do I trust you?' A better question is 'Do I trust you to do x?' We need to think of trust as *trusting someone to do something*. For instance, you can trust me to write an article but trusting me to, say, drive a lorry would be a grave mistake. You can trust me to teach twenty-something-year-old graduate students, but put me in a class of five-year-olds and I would probably lose my rag in my attempt to teach them to read and write. What we are trusting someone to do fundamentally changes the alchemy and order of importance of those four necessary traits: competence, reliability, integrity and benevolence. Trust is highly contextual.

I am not suggesting that we have to go through this kind of

assessment with every person we need to trust. Take something simple like getting on a bus or train. We don't want to be assessing the driver's skills. If we thought about every decision to trust, we would spend the entire day asking questions and making checklists. We might never even leave the house. But let's say we're hiring someone for a job. How do we make good decisions about whether or not to trust what they claim about themselves?

Not so long ago, CVs were the principal tool in job hunting, listing what someone had – allegedly – done and with whom, but they didn't provide much proof. Indeed, a survey done in the UK by Powerchex, a company that screens CVs on behalf of finance companies, found that, out of 4,735 job applications, 18 per cent contained outright false information.[21] The most common lie was to claim a 2:1 university degree when they had been awarded a 2:2. Embellishing the truth, such as exaggerating job titles (from project assistant to project director) and tasks, is very common. CVs were often 'creative', at best, but pretty much all the average employer had to go on.

Today, many people send their LinkedIn page instead of a CV and include links to their other online portfolios and social profiles. Think for a minute about the number of profiles out there that contain information on you. I easily have fifteen: my profiles on Amazon, eBay, LinkedIn, Facebook, BlaBlaCar, Uber, YouTube, Twitter, TED, Penguin author page, my personal website, my literary agent's website, my speaking agents' websites and so the list goes on. And that does not even include all my accounts that require a Facebook login such as Spotify, Airtasker and Airbnb. Online profiles are another example of the shift in trust signals. Information that used to be held by institutions or small groups of friends, family or colleagues is now distributed among many people. In this sense, trust signals have become socially fluid.

In online services such as UrbanSitter, people create specific online profiles listing detailed information. The amount and type of details people will voluntarily disclose is extraordinary. 'One parent included a long description about their pot-bellied pigs,'

Perkins explained. 'They wanted to make sure the sitter was comfortable with the family keeping pigs as pets. It sounds weird but is actually great expectation management.'[22] Both sides knew what they were in for.

The information a sitter provides in their profile is verified through different online checks. Do you really have cardiopulmonary resuscitation (CPR) certification? Did you really get a Level 5 childcare diploma from South Thames College? Do you really have a clean driving record? Or in the case of Doris, do you really belong to the Salvation Army? In fact, only 25 per cent of the sitters that start the registration process make it on to the platform. Put differently, 75 per cent are rejected.[23]

UrbanSitter uses forty different criteria to find the six sitters most compatible to the parent. The algorithm takes into account the age of the children, the parents' social connections to the sitter, where they live, when the sitter is needed, specific preferences and so on. The marketplace also uses a reputation system just like eBay – parents leave feedback and rate sitters after each transaction.

'People rate everything now. Your Uber driver, the guy who delivers your food, and not all things are equal. There are only so many things you can do wrong or right when you're delivering my food,' says Andrea Barrett, UrbanSitter co-founder and vice president of product.[24] 'But when you're watching my child, there are a lot of things that can go awry. Your ability to deal with the behaviour of my child, how much you engage with my child, how much you cleaned up, were you friendly, did I like you? There's a lot going on.'

Reviews significantly influence a sitter's ability to get a booking – a sitter without reviews is two times less likely to get a booking than a sitter that has at least one. A valuable badge experienced sitters and parents carry is called 'Repeat Families'. Basically, it means you were invited back.

Reliability is also easier to demonstrate online. If you have booked a place on Airbnb, you may have noticed that hosts are categorized by how quickly they respond. My response rate is 100 per cent but my response time is, um, twenty-four hours. This means I respond to all new messages but by Airbnb standards I am slow to message back. Similarly, UrbanSitter categorizes sitters based on how quickly they reply. 'If you are looking for someone last minute, it's good to know upfront how long it will take them to respond,' says Perkins. 'But on a deeper level, I think it's an indicator of reliability in some weird way. If this person is really slow to respond, will they show up on time? Are they really interested?' Perkins is right; time is often used as an indicator of reliability.

The hardest trait, without a doubt, both to prove and predict online, is integrity. How do you get a real read of someone's intentions? Perkins knew this was a big problem she had to address. In 2014, she wondered if asking the sitters to make short videos about themselves and their interest in kids would make a difference.

Most videos start with, 'Hi, my name is so and so and I would like to give you a bit of information about . . .' Then they get more personal, explaining why they want to look after your kids. One wondrous offshoot of the videos is seeing the rooms people are in. Some sit casually in their bedrooms on unmade beds, others choose the sofa in tidy living rooms.

I tried an informal experiment. I read the written profiles of twenty sitters and wrote down characteristics that summed up my first impressions. I found I kept using the same vague words such as 'nice' or 'seems friendly'. I then did the same exercise using the videos. The descriptors were clearly more specific, such as 'considerate', 'warm' or 'nervous'. Hearing sitters talk and seeing a small glimpse of their environment felt like seeing inside their lives. 'It allows them to be human,' Perkins says. Today, it is mandatory for all sitters to leave a thirty-to-ninety-second short clip video about themselves. There is this, though: Doris would have made a great video.

So would UrbanSitter have caught that she was a fraud?

On 5 July 1993, a 'dog' cartoon by Peter Steiner appeared in the *New Yorker*. The pen-and-ink artwork features two dogs, one sitting on the floor and the other in a chair in front of a computer. Underneath is the prescient caption, 'On the internet, nobody knows you're a dog.'[25] In 2013, to mark the twentieth anniversary of that iconic cartoon, the web comic 'Joy of Tech' created an updated version. It featured two agents with sunglasses standing in the National Security Agency (NSA) surrounded by computer screens. 'Our Metadata analysis indicates that he is definitely a brown lab. He lives with a white-and-black-spotted beagle mix and I suspect they are humping,' reads the caption.[26]

On a crisp autumn morning, I visit the modest offices of Trooly, an 'Instant Trust Rating' venture launched in July 2014 based in Los Altos, a sleepy backwater city to the north of Silicon Valley. Sitting in the start-up's meeting room, Steiner's cartoon came to mind.

Savi Baveja, the company's CEO, wants to show just how powerful these new trust checks can be. 'So, what do you think of the idea of running you through the Trooly software to see what comes up?' he says, smiling encouragingly.[27]

I blush, trying to recall all the bad or embarrassing things I've ever done. Who knows what my past might reveal? My many speeding and parking tickets? The weird websites I sometimes spend time on (for research purposes, of course)? Old photos from the Piers Gaveston ball I once attended, Oxford University's notorious student shindig that makes the film *Eyes Wide Shut* look like a conservative tea party? I laugh nervously.

'Don't worry – we can project it on to the large screen so you can see what is happening in real time,' Baveja, in his late forties, offers. Somehow I don't find that reassuring.

Anish Das Sarma, the company's chief technology officer and formerly a senior researcher at Google, types my first and last name into the Instant Trust program. Then my email address. That's it. No phone number, age, date of birth, occupation or address.[28]

'Trooly's machine learning software will now mine three sources of public and permissible data,' explains Baveja. 'Firstly, there are public records such as birth and marriage certificates, anti-money-laundering watch lists, the National Sex Offender Registry and so on. Any global register that is public and digitized is available to us.' Then there is a super-focused crawl of the deep web. 'The deep web is still the internet but it's hidden; the pages are not indexed by typical search engines,' he says. So who uses it? 'Hate communities is a big one. Paedophiles. Guns and weapons. It's where the weird people live on the internet.'

The last source is social media such as Facebook and Instagram. Official medical records are off-limits. However, if you tweeted, 'I just had this horrible back surgery,' it could be categorized as legally permissible data and used. Baveja and his team spent nine months weighing up what data they should and should not use. Data on minors was out. 'In some countries,' he says, 'there is a legally agreed definition of the difference between private information and sensitive private information – the latter includes medical, plus race, religion, union membership, etc. The latter is what we drew the line on, as we were very aware of the creepy factor.'

After about thirty seconds, my results appear. 'Look, you are a one!' says Baveja. Profiles are ranked from one to five. 'Only approximately 15 per cent of the population are a one; they are our "super-goods".'

I feel relief and a tinge of pride. There must be a lot of people in the middle, but how many fall into the category of 'super-bad'?

'About 1 to 2 per cent of the population across the countries Trooly covers, including the US and the UK, end up between five and four,' Baveja explains. 'There are some exceptions in some populations we've done because some populations attract super creeps.'

Baveja was previously a partner at the consulting firm Bain & Company. One of his longest-standing clients was a 'well-known online marketplace'. It started Baveja thinking about the importance of trust in the digital world. 'Our client needed 6 per cent of their entire budget – hundreds of millions of dollars – to respond to things going wrong in their marketplace,' he tells me. 'It got me thinking how the typical star-rating system was not adequate to prevent a very large number of incidents online.'

When you first connect with people, total strangers living on the other side of the world, how do you know if they pose a serious risk? 'People talk a lot about reviews and ratings. Well, by definition, they're backward-looking,' says Baveja. Indeed, the goal of Trooly is to fill the 'trust gap' caused by the sheer speed of online commerce.

Meanwhile, Baveja's wife was running a small dental practice. People would refuse to pay or threaten to leave bad reviews. And at the weekend, there would be numerous callers demanding drugs. 'It occurred to me that small businesses, relative to big businesses, know very little about their customers,' says Bajeva. 'Wouldn't it be cool if small businesses had a way of weeding out potentially bad customers?'

To get my trust score, Trooly's software crawled more than 3 billion pages of the internet, from around 8,000 sites, in less than thirty seconds. The data was consolidated into three buckets. The most basic was verifying my identity. Was I who I claimed to be? This is easily done by scanning, say, my personal website against my university profile. Next was screening for unlawful, risky or fraudulent activity. But it's the third category that is fascinating. I was assessed against the 'dark triad', a trio of callous personality traits that make con artists tick: narcissism (selfishness with excessive craving of attention), psychopathy (lack of empathy or remorse) and Machiavellianism (a highly manipulative nature with a lack of morality). Unfortunately, Baveja can't give me a separate score here, but it's safe to say I passed.

Trooly was awarded a US patent two years ago for this software, 'determining trustworthiness and compatibility of a person'. Its algorithm was also programmed to take into account the 'big five' traits – openness, conscientiousness, extraversion, agreeableness and neuroticism – widely accepted by researchers in the 1980s as a key way to assess personalities. 'Trooly developed sophisticated models to predict these traits using hundreds of features from an individual's online footprint,' says Baveja. 'It was super-interesting figuring out what, in an online footprint, might help you predict if someone is going to be, say, super-neurotic or rude. If you look at someone's Twitter account and it's peppered with excessive self-reference and swear words, the person is much more likely to be antisocial.'

Baveja is a warm, considerate and clearly intelligent man. His sentences tend to contain lots of thoughtful questions, as if he is constantly seeking better answers. He went to Stanford University for his undergraduate degree in electrical engineering. A few years later, he completed his MBA at Harvard Business School where he graduated as a Baker Scholar, meaning he was one of the top students in his class. He then became a management consultant, rising to the highest echelons of Bain & Company. 'While at Bain, I started to look into traditional background and credit checks, and realized how dangerously flawed they are,' he says. 'A background check is just that – it is retrospective and does not foresee the future. But does this have to be the case? I thought there must be something better.'

Our behaviours have changed but the trust mechanisms we use in society have stayed pretty much the same. For one thing, in the US, UK and much of Europe, the current background-check system is still slow and manual, often relying on low-paid and overworked court runners to rummage through records and so on. No wonder all kinds of mistakes happen, especially if your last name happens to be Jones, Smith or Harris, or another common moniker. Ron Peterson, who lives in California, knows this problem all too well. 'In Florida, I'm a female prostitute (named Ronnie); in Texas, I'm currently incarcerated for manslaughter,' says Peterson. 'In New Mexico, I'm a dealer of stolen goods. Oregon has me as a witness tamperer. And in Nevada – this is my favorite – I'm a registered sex offender.'[29] If you are falsely mismatched with someone else's felony, it's known as a *false positive*. It's an alarmingly common problem.

Out of the people the traditional checks label 'bad', how many are actually criminals? Worryingly, a study conducted in 2016 by Simone Ispa-Landa, an assistant professor at Northwestern University, and Charles Loeffler, a professor at the University of Pennsylvania, found that one out of three American adults has been involved with the criminal justice system and has a record, even if they were not found guilty of a crime.[30] The United States Attorney General's office recently found something similar. Half of all case files in the background system contain no information about how the cases turned out – whether the person was found guilty or not, or even prosecuted. Bad data fouls the system and the most common groups falsely labelled are blacks and Latinos. 'What we've ended up doing is taking all the biases and pre-existing preconceptions of the criminal justice system,' Baveja says, 'and ossifying them, institutionalizing them in more and more and more decisions where they don't belong.'

Baveja raises question after question about the checks system. 'I can tell you that 50 per cent of what you will find in a background check is going to be either traffic violations or drug convictions,' he says. 'But if I smoked a joint seven years ago, what does it really tell you about me? Does it mean I'm going to be a bad tenant?'

'But surely, if I had been convicted of shooting someone, the check would flag this up?' I ask.

'No, a fail is a fail in a traditional background check. It's just not precise,' says Baveja.

But there is another problem: the system can miss people who really are criminals. It's known as a *false negative*. Of the millions of checks done every year, between 1 and 2 per cent turn up a problem. Indeed, the vast majority of people who have done bad things pass their background checks. 'How on earth did we get to a point where we rely on flawed data and processes that lack rigour to determine whether someone should get a job or will be a good tenant?' says Baveja. 'I mean, these are serious decisions, right?'

In June 2017, Airbnb acquired Trooly for an undisclosed, but presumably sizeable, sum. It's part of its investment in troubleshooting, along with people like Nick Shapiro, who was a deputy chief of staff at the CIA until May 2015. A couple of months later, he joined Airbnb as global head of trust and risk management.

'Earning, keeping and facilitating trust has always been and always will be a core part of any functional society,' he says. 'What has changed is where and how trust is exchanged. Rather than placing trust in major institutions like big business, the media or government, people are trusting each other more and more.'

Part of his brief is to figure out how Airbnb responds when things go awry. Take someone like David Carter, who booked a luxury New York apartment for the weekend in March 2014. He told the host, Ari Teman, he was looking for a place for his brother and sister-in-law to stay while they were in town for a wedding. In fact, his 'in-laws' turned out to be guests for a rowdy orgy featuring a 'big beautiful woman' and kinky things with stuffed animals. Teman only discovered the X-rated soirée after he popped back into his building to pick up his luggage.

Carter had a verified Airbnb account and positive reviews. 'We have a responsibility to go back and see what, if anything, we could have done differently,' Shapiro says. 'Was there a mistake by someone?' In other words, how could the Carters of the world be weeded out?

'You could have easily figured out that something like that was going to happen,' says Baveja. 'You didn't need to go deep into Carter's psyche to work out this guy was a professional organizer of sex parties.' A simple Google search on Carter's email address led to online advertisements for events such as 'Turn Up Part 2: The Pantie Raid' and 'BBW Panty Raid Party'. One person even gave the apartment's address in a tweet for an 'XXX FREAK FEST'.

'I learned new acronyms looking at his profile,' says Baveja. 'I'm blushing just thinking about it.'

I tell Baveja and his team about Doris. Would the search have caught her if she had applied to UrbanSitter? The answer is a definite 'yes'. My parents would have known that she didn't belong to the Salvation Army, had no previous childcare experience and had a patchy criminal history. Doris would not have made the cut.

Should we embrace these new trust algorithms? Baveja and Shapiro speak openly about the immense responsibilities that come with trying to take what are essentially ethical decisions and translate them into code. How much of our personal information do we want trawled through in this way? And just how comfortable are we with letting an algorithm judge who is trustworthy?

During my test at Trooly, I found myself worrying about tiny or long-ago 'transgressions' being held against me. Do companies take note of those?

'I don't think anyone likes to be judged, whether by a robot or another person, but that isn't what our screening is about,' Shapiro insists. 'We are looking for major risks such as a member of a hate group, a violent criminal past or a fake identity. We don't care if you sent a stupid tweet or got a parking ticket last week.'

There are other questions. What, for example, are the consequences for 'digital ghosts'? People like my husband, who has never used Twitter, would not know how to log in to Facebook, doesn't have a profile on LinkedIn. Does his 'thin file' reduce his ability to be considered trustworthy?

'For around 10 to 15 per cent of people we screen, we can't generate a confident trust score,' Baveja admits. 'There's either not enough of a digital footprint out there or enough accurate inputs.'

On sites like Airbnb, invisibility will not necessarily work against you. 'We are looking for derogatory information,' says Shapiro, 'and the absence of information is not seen as a negative in itself. Not having a large digital footprint isn't something we count against people.'

Online trust is set to get faster, smarter and more pervasive. Can technology strip out all the risk of dealing with strangers?

'No way,' says UrbanSitter's Perkins. Humans are complex moral beings, capable of nuanced thought, instincts and common sense, and it would be foolish to remove ourselves from the picture entirely. 'If a sitter shows up at your door and you get a weird feeling, it doesn't matter if they have passed checks, how well reviewed they were or what you thought about them online, say you are suddenly not feeling well and cancel. Go with your gut.'

It's complicated. Our intuition may be strong but sometimes we tune in to trust signals that are loud even though they are not in fact good indicators of trustworthiness. Doris, with her spectacles and Scottish accent, is a prime example.

Making the odd mistake when it comes to trust, taking a leap, is sometimes how we open up new possibilities or find ourselves in unexpected situations, both exciting and dangerous. It's how we place our faith in strangers, without knowing what might come of it.

At the end of the day, the onus is still on us to decide where to place our trust and in whom, but we are now in a much better position to ask the right questions and find the right information. We are, as a result, smarter than ever before. That should mean we're less likely to hire a narco bank robber as a nanny.

6

Reputation is Everything, Even in the Dark

Depending on how strictly you want to define the terms of an 'online trade', you could claim that the first thing to be bought and sold on the internet wasn't a CD or pizza, it was a small bag of weed. In the early 1970s, a group of students at Stanford University and Massachusetts Institute of Technology made an online drug deal via ARPANET, the precursor to the internet we know today.[1] Since then, it has become remarkably fast and easy to score drugs online, through the so-called 'darknet'.

You can't get to the darknet using your regular web browser; it can only be accessed via an anonymizing software called Tor (an acronym for 'The Onion Router'). Instead of a web address ending in a .com or .org, darknet URLs are a hash of random letters and numbers that end in .onion. Originally developed by the US Naval Research Laboratory for the purpose of protecting government communications, Tor has become a handy privacy tool for journalists and human rights organizations that need to mask their browsing activity and hide their identity and location. Of course, its subterranean nature means it also attracts criminals who can exchange drugs and other illegal goods, from guns to child pornography, online and in relative obscurity.

Stumbling into the darknet is like stumbling into a shadowy and mysterious parallel universe, where everything looks oddly the same as on the regular web – strangely familiar – except that its consumer sites are selling AK-47s or counterfeit passports, instead of pre-loved Hermes handbags and Jamie Oliver cookbooks. It means a visitor doesn't need to be a hacker or computer whiz to navigate it; it's remarkably easy to find and buy illegal goods and services. Google doesn't search onion sites but Grams does (FYI, the address for Grams is: grams7enufi7jmdl.onion) and its site looks incredibly similar, from the brightly coloured rainbow logo to the 'I Feel Lucky' button.

Let's say you type in 'ecstasy' on Grams. The search engine trawls cryptomarkets such as BlackBank, Mr Nice, Pandora and SilkRoad4 (now on its fourth life) and provides a list of results showing the name of the seller, price of the product, a brief description and the exact URL. As the creator of Grams told *WIRED* magazine in an anonymous interview: 'I noticed on the forums and Reddit, people were constantly asking "who had the best product X and was reliable and not a scam?" I wanted to make it easy for people to find things they wanted on the darknet and figure out who was a trustworthy vendor.'[2]

I had read a lot about darknet drug sites but I was still gobsmacked at just how much they look like conventional e-retailers such as Amazon. They would appear reassuringly familiar to any online shopper. There's even the usual amount of competition and cornucopia of choice. It's just that the listings, row after row, are for cocaine, blotter (LSD), ecstasy, opioids, dimethyltryptamine (DMT), heroin, hash, cannabis and almost any other drug a user could possibly want. With just a few clicks, buyers can browse a mind-boggling selection, pay for the drugs in the traceless digital currency bitcoin (unattached to any central bank) and have them delivered unknowingly by the postman.

In October 2013, the darknet achieved notoriety when an illicit drug site called Silk Road was shut down by the FBI. The site owner and administrator, twenty-nine-year-old Ross William Ulbricht, aka Dread Pirate Roberts (DPR), was arrested at a public library in San Francisco. Convicted of money laundering, computer hacking and conspiracy to traffic narcotics, he has since been sentenced to life in prison. Appropriately, Ulbricht took his fictional namesake, Dread Pirate Roberts, from a character in the book and film *The Princess Bride.* In the story, Roberts is not one man but one of many who pass on the name, reputation and pirate business from one person to the next. At the time of the site's closure, the FBI estimated that Silk Road had 13,000 drug listings and had processed approximately $1.2-billion-worth of sales.[3] DPR was believed to be making an estimated $20,000 a day in the 6–12 per cent commission the site charged on every transaction.[4]

Ulbricht was a university-educated guy who grew up in Austin, Texas. A self-proclaimed idealist, he believed drug use was a personal choice. He was essentially a twenty-first-century digital libertarian. In Silk Road's code of conduct it stated, 'We refuse to sell or list anything the purpose of which is to harm or defraud another person.' DPR also wrote, 'Treat others as you wish to be treated.' Ulbricht wanted to create a trusted trading ground where people could buy and sell drugs free from violence and the reach of government laws. Critically, DPR was not just revered – vendors and buyers on Silk Road trusted him. But once he was arrested, their belief was shaken in the safety of the whole system. It wasn't enough, however, to bring it crashing down. And other sites sprang up to replace Silk Road.

How are visitors persuaded to trust these sites? Shared control, for one thing. When a buyer places an order, the bitcoin goes into an escrow account and is only released to the vendor once the order has been confirmed as received. (It's a similar escrow system to the one used on Alibaba.) The buyer, seller and the site's administrators control the account; two out of the three must sign off on the deal before the bitcoin can be moved. It's designed to make it far more difficult for vendors and buyers to scam one another. On the whole, it works, although, like most systems, there are pitfalls and bad apples. For example, unscrupulous administrators can shut down a marketplace and make off at any time with all of their users' bitcoins escrowed in the site. And some do. It's known as an *exit scam*.

In March 2015 the popular Silk Road descendant Evolution mysteriously disappeared overnight. It had nothing to do with a government bust; just simple greed. The site's administrators, Verto and Kimble, emptied the market's bitcoin coffers and ran off with an estimated $12 million.[5] A few days after the incident, a user named NSWGreat, who had previously self-described herself as an Evolution 'public relations' staffer, put a post on Reddit's darknet markets forum.

'I am so sorry, but Verto and Kimble have f—ked us all. I have over $20,000 in escrow myself from sales,' NSWGreat wrote. 'I'm sorry for everyone's losses, I'm gutted and speechless. I feel so betrayed.'[6]

Soon after Evolution shut down, lesser-known markets like Abraxas, Amazon Dark, Blackbank and Middle Earth also subsequently disappeared for unknown reasons, but it's assumed they were also exit scams. The utopian ideals of the darknet – 'us against the government and over-regulation' – seemed to have had their heyday. And, of course, vendors and buyers who have been ripped off have little recourse, beyond warning each other. They can hardly go to the ombudsman and complain their LSD didn't turn up and demand their bitcoins back. How much damage did that do to trust in the darknet?

Dr James Martin is a renowned expert on cryptomarkets. He's a professor at Macquarie University and author of *Drugs on the Dark Net*. I interviewed Martin from his flat in Melbourne just after he had come back from, appropriately, Amsterdam. He is particularly fascinated by how cryptomarkets use technology to allow people to communicate and engage in new forms of self-governance outside state and traditional regulatory control.

'What interests me is the way technology could be used to transform an illicit drugs trade that has been debased and corrupted by four decades of the global War on Drugs,' he says.[7] 'The fact is, you have got thousands of people – drug users and drug retailers – who have been stereotyped as "untrustworthy", who don't have a reputation for being the most reliable people. So how do they create highly functioning markets that are non-violent and self-regulate based on trust?'

Martin thinks law enforcement agencies and the police have been surprised by the darknet. '[They've seen] how criminals are creating peaceful communities selling dangerous products; communities that work really well the vast majority of the time.'

Yet haven't the exit scams now damaged that trust in the darknet? 'There isn't a trust crisis but people are definitely more sceptical,' he says. 'Scams undermined the faith in the whole system. It wasn't the "big bad Drug Enforcement Agency (DEA)" that was destroying the markets; it was people from the inside. That really chipped away the idealism and shook people's faith in the darknet community.'

Martin likens it to research one of his colleagues carried out looking at desertion within armies and how it affected the remaining troops. The study found that if fellow soldiers were killed by the enemy, they were seen to be doing their job and it stiffened the resolve of those left. If they fled voluntarily, however, it undermined faith in what they were all fighting for, faith in the whole system. 'That,' says Martin, 'is what we saw with the exit scams.'

Even so, the scams didn't deliver a mortal blow. 'Vendors and customers have just moved past them and repopulated other sites,' Martin says. It's like a game of whack-a-mole: as soon as one site closes down, another pops up.

Despite the Silk Road bust and others since, the drug business on the darknet, like other forms of e-commerce, is thriving. According to the 2016 Global Drug Survey, approximately 22 per cent of UK drug users have sourced and bought their drugs online.[8] Globally, almost one in ten participants reported having bought drugs off the darknet. Significantly, 5 per cent of respondents stated that they had not bought drugs before purchasing them through darknet markets.[9] The darknet makes drug-buying easier but also seem less risky to some people.

An in-depth study by the European think tank Rand found that the number of transactions on illegal drug sites has tripled since 2013.[10] The United Kingdom is the largest online drug market outside the United States, with darknet sellers doing close to 21,000 drug deals a month. British vendors took home on average around £5,200 per month. But the most successful darknet dealers were making upwards of £200,000.[11]

The truly fascinating thing, however, about drug marketplaces is not how many there are or how much dealers make but just how well they work. While we should rightfully fear that these sites could lead to higher levels of drug use, it's hard not to be impressed by their apparent quality control and efficiency. People, ordinary people, transact with each other globally, at high speed and in great volume. The darknet can teach us a lesson or two about trust.

In street dealing, selling is often restricted to customers a dealer already knows. New customers tend to be introduced to dealers only by a broker, a trusted intermediary. 'My mate is not a cop, you can sell her drugs and she won't dob you in,' is how a broker might vouch for someone. Trust is crucial in conventional drug-dealing but it's small-scale, local trust, meaning the trust lies directly between a few people.

In contrast, the darknet is an open network. Choice isn't limited by whom you know or where a dealer lives. 'Cryptomarkets represent a kind of super broker. They are able to facilitate contact between many, many more vendors and customers than any individual possibly could,' says Martin. 'But the trust in this instance is represented in quantitative metrics, such as reviews and ratings that substitute for the personal trust that has been historically critical for the drug trade. And that is transformative.'

Trust used to be a very personal thing: you went on the recommendations of your friends or friends of friends. By finding ways to extend that circle of trust exponentially, technology is expanding markets and possibilities. In the case of the darknet, it is creating trust between the unlikeliest of characters, despite a heavy cloak of anonymity.

The darknet is peopled by hundreds of thousands of drug users and vendors who would commonly be stereotyped as untrustworthy, the worst of the worst, yet here they are creating highly efficient markets. Effectively, they are creating trust in a zero-trust environment. Nobody meets in person. There are obviously no legal regulations governing the exchanges. It looks like a place where buyers could get ripped off. Theoretically, it would be easy for dealers to send lower quality drugs or not deliver the goods at all. Yet this rarely happens on the darknet and, overall, you're more likely to find buyers singing hymns of praise about the quality of the drugs and reliability of the service.

An extensive report published in February 2016 by the European Monitoring Centre for Drugs and Drug Addiction (EMCDDA) revealed that drugs available through darknet markets tend to be of a higher purity than those available on the streets.[12] Similarly, a study by Energy Control, a Spanish think tank, also confirmed a quality premium in darknet drugs. Volunteers were asked to send random samples of drugs they had bought from online sites and offline dealers for testing. More than half of the darknet cocaine samples contained nothing but cocaine, compared with just 14 per cent of those bought on the Spanish streets. Even the FBI testified in the Silk Road hearings that, of more than a hundred purchases of drugs they made online before the closing of the site, all showed 'high purity levels'.

It seems dealers are more honest online. And that tells us something about how people create trust in this criminal ecosystem, with the help – or supervision – technology can provide. Given the nature of its wares, the darknet may seem like an alien and subterranean world – and some of it undoubtedly is – but at its core, it's about people connecting with other people. It's just another incarnation of the new kind of trust-building that technology has facilitated. The same dynamics, the same principles in building digital human relationships, apply. In that sense, it's almost comically conventional.

Turns out, drug dealers care about their online brand and reputation and customer satisfaction as much as Airbnb hosts or eBay sellers. A typical vendor's page will be littered with information, including: how many transactions they have completed; when the vendor registered; when the vendor last logged in; and their all-important pseudonym. It will also feature a short description about why a user should buy drugs from them, refund policy information, postage options and 'stealth' methods (measures used to conceal drugs in the post). Even if you're not in the market for what they're selling, it's hard not to be impressed; vendors put in real effort to demonstrate their trustworthiness.

Traditionally, the words 'drug dealer' bring to mind some thuggish or shady lowlife, a badly educated tough swaggering around in a leather jacket or lurking threateningly on a street corner. Someone you don't want to mess with. True or not, that persona of a heartless intimidator doesn't work on cryptomarkets. There, vendors need to project a cleaner image. Some even display specific logos and taglines and their brand message is loud and clear: 'We care about you' or 'Your satisfaction is our priority'.[13]

Indeed, marketing strategies used on the darknet look remarkably like standard ones. There are bulk discounts, loyalty programmes, two-for-one specials, free extras for loyal customers and even refund guarantees for dissatisfied punters. It's common to see marketing techniques such as 'Limited Stock' or 'Offer ends Friday' to help boost sales. While reviewing these deals, I had to remind myself I was looking at an illegal drugs marketplace, not shopping for shoes on Zappos.

Some vendors, eager to build brand, label their drugs as 'fair trade' or 'organic' to appeal to 'ethical' interests. 'Conflict-free' sources of supply are also available for the benevolent drug taker. 'This is the best opium you will try, by purchasing this you are supporting local farmers in the hills of Guatemala and you are not financing violent drug cartels,' promised a seller on Evolution (before it was shut down).[14]

New vendors will offer free samples and price-match guarantees to establish their reputation. Promotional campaigns are rife on 20 April, also known as Pot Day, the darknet's equivalent of Black Friday. (The date of Pot Day comes from the North American slang term for smoking cannabis which is 4/20.) 'It's not anonymity, Bitcoins or encryption that ensure the future success of darknet markets,' writes Jamie Bartlett, author of *The Dark Net*. 'The real secret of Silk Road is great customer service.'[15]

After a buyer receives their drugs, they are prompted to leave a star rating out of five. Nicolas Christin, an associate research professor at Carnegie Mellon University, analysed ratings from 184,804 pieces of feedback that were left on Silk Road over the course of an eight-month period. On the site, 97.8 per cent of reviews were positive, scoring a four or five. In contrast, only 1.4 per cent were negative, rating one or two on the same scale.[16] Just how much can people trust those good reviews?

Some observers suspect the darknet suffers from *review inflation*. Similar studies have been conducted on conventional marketplace rating systems and found that feedback is also overwhelmingly positive. For instance, on eBay less than 2 per cent of all feedback left is negative or neutral. One explanation is that dissatisfied customers are substantially less likely to give feedback.[17] It means the most important information, the negative reputation data, is not being captured.

Social pressure encourages us to leave high scores in public forums. If you have experienced an Uber driver saying at the end of a trip, 'You give me five stars, I'll give you five stars,' that's tit for tat or grade inflation in action. I know I'm reluctant to give a driver a rating lower than four stars even if I have sat white-knuckled during the ride as he whizzed through lights and cut corners. Perhaps that is why Jason Dalton was a 4.7. Drivers get kicked off the Uber platform if their rating dips below 4.3 and I don't want to be responsible for them losing, in some instances, their livelihood. Maybe they are just having a bad day. That, and the driver knows where I live. In other words, reviews spring from a complex web of fear and hope. Whether we are using our real name or a pseudonym, we fear retaliation and also hope our niceties will be reciprocated.

It would be easy to conclude, then, that the ratings we rely on to make assessments are often not an accurate reflection of the experience. But they can still make us more accountable to one another. For instance, I sometimes drop towels on the bathroom floor when I am staying in a hotel. But I would never do this as a guest staying in someone's home on Airbnb. Why? Because I know the host will rate me, and that rating is likely to have an impact on my booking requests being accepted by other Airbnb hosts in the future. It illustrates how online trust mechanisms will have on impact on our real-world behaviours in ways we can't yet even imagine.

While ratings might be exaggerated, or even fabricated, some sites are taking steps to reduce the problem of positive bias. For instance, in 2014 Airbnb introduced a *double-blind* process where guest and host reviews would only be revealed after both are submitted, or after a fourteen-day waiting period, whichever came first. The result was a 7 per cent increase in review rates and a 2 per cent increase in negative reviews. 'These may not sound like big numbers but the results are compounding over time,' says Judd Antin, director of research at Airbnb. 'It was a simple tweak that has improved the travel experiences of millions of people since.'[18] Like any game, it's about figuring out the rules that put downward pressure on an unwanted behaviour until it doesn't exist any more.

There is, however, another way to look at the 97.8 per cent of positive reviews on darknet sites. Perhaps they are an accurate reflection of a market functioning remarkably well most of the time, with content customers. Even if review systems are not perfect, and bias is inevitable, it seems that they still do their job as an accountability mechanism of social control. Put simply, they make people behave better.

A question I have often asked myself is whether limiting people to a score or star rating is really that helpful. Many marketplaces are now asking people to rate against a particular trait that is more relevant to trust in a specific context. For instance, on BlaBlaCar, people are rated from one to five on their driving skills. On Airbnb, hosts are rated from one to five on their cleanliness, accuracy, value, communication, arrival and location.[19] On Uber, when riders give a low or high rating, you must specify the reason for doing so, such as cleanliness, driving, customer service or directions. On drug marketplaces, how clean a dealer is and how well they drive is irrelevant. What would be the trust measures there?

Turns out, ability and character also applies to drug vendors. To highlight reliability, many reviews point out the speed of response and delivery. For example, 'I ordered 11.30 a.m. yesterday and my package was in my mailbox in literally twenty-five hours. I'll definitely be back for more in the future,' commented a buyer on Silk Road 2.0. Other reviews focus on the product quality: 'Amazing weed. Fast shipping. Packaging very secure. Took me a bit long to get in the double-seal vacuum seal but well worth it. Would highly recommend.' One of the ways skills and knowledge are reviewed is how good a vendor's 'stealth' is, that is, how cleverly they disguise their product so that it doesn't get detected. 'Stealth was so good it almost fooled me,' wrote a satisfied buyer on an MDMA listing on the AlphaBay market. Established vendors are very good at making it look (and smell) like any old regular package. Excessive tape or postage, reused boxes, presence of odour, crappy handwritten addresses, use of a common receiver alias such as 'John Smith' and even spelling errors are bad stealth.

There is a clear incentive for vendors consistently to provide the product and service they promise: the dealers with the best reviews rise to the top. No feedback, either negative or positive, can be deleted, so there is a permanent record of how someone has behaved. Just like the Maghribi traders and Alibaba sellers discovered, past behaviour is used to predict future behaviour. 'The future can therefore cast a shadow back upon the present,' Robert Axelrod wrote in *The Evolution of Cooperation*.[20] Or to put it another way, vendors have a vested interest in keeping their noses clean right from the word go.

Reputation is trust's closest sibling; the overall opinion of what people think of you. It's the opinion others have formed based on past experiences and built up over days, months, sometimes years. In that sense, reputation, good or bad, is a measure of trustworthiness. It helps customers choose between different options and, with luck, make better choices. It encourages sellers to be trustworthy, in order to build that reputation, and it weeds out those who aren't.

It isn't quite that simple, however. Price is also a factor in the value of reputation, and reputation can influence price.

Consider this scenario: two vendors are both offering Purple Haze cannabis on exactly the same shipping and refund terms. One vendor, let's call him BlazeKing, has only three reviews, with an average score of 4, and is offering the cannabis for $12.50/gram. The other vendor, CandyMan2, has fifty-two reviews, with an average score of 4.9, and is offering the same Purple Haze but for a slightly higher price at $12.95/gram. Who to go with? Obviously, the majority of customers would choose CandyMan2. The reviews reduce uncertainty for the buyer.

CandyMan2's higher reputation comes with a price to the customer, though not a huge one. The value of the reduced risk is $0.45/gram. In this sense, online reputation is functioning as a risk premium.[21]

Now let's say the choice is between BlazeKing, still with three reviews and cannabis for sale at $12.50/gram, and another vendor, FlyingDynamite, also with fifty-two reviews and an average score of 4.9, but selling the same drugs for $16.50/gram. Now the risk premium for a vendor with a higher reputation is $4/gram. Is the vendor's reputation worth the significant difference in price? For some, it may be; for others, not. Even reputation, while valuable, has a price ceiling. There is only so much more a customer is willing to pay for goods or services based on a seller's high reputation.

Still, I was intrigued how influential a vendor's reputation really is on cryptomarkets. So I tracked down a buyer, let's call him Alex. He describes himself as a 'casual drug taker' who likes to smoke a lot of weed and occasionally take ecstasy at the weekend. Towards the end of 2014, Alex switched from buying drugs from a dealer who lived relatively close by, to buying through cryptomarkets. What made him switch? His answer confirms everything we've just explored. 'Rather than buying drugs from a friend of a friend of a guy I met in a bar, I can buy drugs after reading dozens of reviews of their service,' he explains. 'I feel confident I am getting exactly what I am paying for.' It echoes what drug surveys indicate: 60–65 per cent of respondents say that the existence of ratings is *the* motivation for using darknet marketplaces.[22]

And that brings us back to the matter of feedback and reputation systems on the darknet. Those systems make both sides more accountable but they are not infallible. Ratings are gamed in a similar way to other sites such as Amazon, Yelp and TripAdvisor. A common trick is a practice known as *padding* feedback. Vendors essentially purchase their own drugs from a series of fake buyer accounts they have created. The glowing reviews look legitimate when they have in fact been created by the vendor. It's the online equivalent of stuffing the ballot box. Politicians do it. Advertisers do it. TripAdvisor is bedevilled by it. And drug dealers do it.

An industry of darknet 'marketing' services has sprung up, peopled by fibbers and promoters who are willing to create rave reviews and posts to help boost a vendor's reputation. 'Hi, my name is Mr420 and we started out and still are a small group of college public relations majors,' wrote a darknet PR vendor. 'We would be interested in keeping your product, thread or listing at the top of the forums.' Fake reviewers on Amazon will get free books, and hotels will often offer discounts for a positive review on TripAdvisor. The likes of Mr420 will pad feedback in exchange for free drug samples. Vendors regard it simply as brand management, doing something to make themselves look better.

It's the same practice Amazon took a suit against in a landmark reputation case. On 16 October 2015, in Washington, DC, the company sued 1,114 individuals for selling positive five-star reviews to Amazon sellers and Kindle authors.[23] All the defendants in the case were advertising their services on Fiverr, an online marketplace where freelancers offer to do minor tasks for a flat rate of $5.

It might be hard to see how Amazon could suffer a loss of revenue from dodgy reviews; products with high-star ratings sell more, right? But Amazon was smart enough to know it needed to crack down on fake reviews because they undermine the foundations of trust in online marketplaces. If reviewers and their reviews can't be trusted, the whole system falls.

But where is the line? Say I send this book to a hundred friends and colleagues, and ask them to leave a nice review on Amazon to help boost sales. Is that gaming the system or just common-sense marketing? As the adage goes, fake it until you make it.

'The weird thing is that even though there is a certain amount of gaming, there is an acceptance that it will take place and it's okay,' James Martin says. 'New vendors will tell you that to break into the market they have to generate false reviews themselves. If you have no feedback in your vendor profile, you are not an attractive proposition. So fake reviews get the ball rolling for new entrants.' It's simply the way markets function; a feedback system will never be perfect.

Gaming the feedback system is also used by rival vendors who want to gain market share. In the case of drug sales, how do dealers compete with one another when violence and turf wars are no longer an option? They engage in online wars. One tactic is sock puppetry (or 'socking'), where a rival hides behind an online identity in order to tarnish the reputation of a competitor. It's a common behaviour, even esteemed professors do it; quite badly, in some cases.

Take Professor Orlando Figes, a critically acclaimed and prize-winning author, who has written eight books. In 2010, he was caught posting damaging critiques of his rivals' books on Amazon.[24] Under the aliases 'orlando-birkbeck' and 'Historian' (drug dealers could teach him a thing or two about selecting a better pseudonym), he called his competitors' works 'dense', 'pretentious' and 'the sort of book that makes you wonder why it was ever published'. Stupidly, he used the same pseudonyms to praise his own work in illustrious detail: 'Beautifully written . . . leaves the reader awed, humbled yet uplifted . . . a gift to us all,' Historian wrote.

When the scandal went public, the professor initially, and somewhat ungallantly, blamed all the reviews on his wife, the barrister Stephanie Palmer. It did not help the situation.[25] Figes ended up paying damages to rival historians Dr Rachel Polonsky and Robert Service, whose work he had slated. In an apology statement he issued to the media, Figes said, 'Some of the reviews were small-minded and ungenerous, but they were not intended to harm.'[26] The same purposeful sniping happens on the darknet. Vendors pretend to be unsatisfied buyers and leave bad reviews.

But gaming can only go so far. Indeed, machine learning systems are already being developed to identify and weed out deceptive reviews. A team of researchers at Cornell University has developed software that can detect *review spam*. In a test on 800 reviews of Chicago hotels on TripAdvisor, the program was able to pick out the deceptive reviews with almost 90 per cent accuracy.[27] In contrast, Cornell's human subjects only managed to pick the fakes about 50 per cent of the time.

It turns out people are beautifully predictable when writing fictional reviews, using similar syntax, language, grammar, punctuation, too many long words and even similar spelling mistakes. The Cornell researchers found that deceivers use more verbs and long words than truth tellers, while the genuine reviewers used more nouns and punctuation.

No doubt these types of review filter will become increasingly sophisticated and commonplace, so we will be better able to trust a review is legit. But there is another simple solution when it comes to countering deceptive reviews: word of mouth. An upstanding community may not be something we associate with the drug trade but darknet markets have a strong sense of community with clear norms, rules and cultures. Users frequently chat to each other on discussion forums such as the DarkNetMarkets on Reddit, publicly calling out dodgy vendors. 'I was looking at this vendor a few hours ago and they had zero feedback. Now they have a bunch,' wrote one user. Customers who continually ask for refunds, claiming that their goods did not show up, are also likely to be shamed.

There are also websites such as DarkNet Market Avengers (dnmavengeradt4vo.onion), that use trained chemists to do random testing of darknet drugs. Users send samples of their drugs to

Energy Control, a drug-testing lab funded by community donations. It tests the products and sends the results back to the user. For instance, if LSD is found to be 'under-dosed' or heroin is found to contain something dangerous like Carfentanil (an extremely potent synthetic opioid which can be life-threatening), the results are posted on the DNM Avengers site, including details of the specific vendor who sold the product.

The result is that fraudsters on both sides of the market are relatively quickly outed and driven out. As James Martin beautifully puts it, 'The darknet is really not dark. Thousands of people hold torches to shine the light on how other people behave. You no longer have to rely on one person but the collective judgements of the entire darknet community.'

Within the next five years, darknet sites could be to street drug dealers what Amazon is to local booksellers or Airbnb is to hotels, even if they do raise different and serious ethical questions. On the one hand, cryptomarkets mean that drugs will be more easily available to more people, which cannot be a good thing. On the other, they reduce the length of the supply chain and some of the risks and criminal behaviour associated with conventional drug-dealing.

Either way, the systems work because customers become enfranchised in them. Technology empowers customers to hold vendors to account and, ultimately, it is only trustworthy vendors who will survive. E-commerce is e-commerce and, even on the darknet, reputation is everything.

7
Rated: Would Your Life Get a Good Trust Score?

On 14 June 2014, the State Council of China published an ominous-sounding document called 'Planning Outline for the Construction of a Social Credit System'.[1] In the way of Chinese policy documents, it was a lengthy and rather dry affair but it contained a radical idea. What if there was a national *trust score* that rated the kind of citizen you were?

Imagine a world where many of your daily activities were constantly monitored and evaluated – what you buy at the store and online, where you are at any given time, who your friends are and how you interact with them, how many hours you spend watching content or playing video games, and what bills and taxes you pay (or not). It's not hard to picture, because most of that already happens, thanks to all those data-collecting behemoths like Google, Facebook and Instagram, or health-tracking apps such as Fitbit that can capture your moves and location at any given time. But now imagine a system where all these behaviours are rated as either positive or negative and distilled into a single number, according to rules set by the government. That would create your *Citizen Score* and it would tell everyone whether or not you were trustworthy. Plus, your rating would be publicly ranked against that of the entire population and used to determine your eligibility for a mortgage or a job, where your children can go to school – or even just your chances of getting a date.

A futuristic vision of Big Brother out of control? No, it's already getting underway in China where the government is developing a system to rate the trustworthiness of its 1.3 billion citizens. I'm sure

George Orwell has rolled over in his grave a couple of times in recent years, but this idea, called the 'Social Credit System' (SCS), must have him doing frantic 360-degree turns in his coffin.

The Chinese government is pitching the system as a desirable way to measure and enhance 'trust' nationwide and to build a culture of 'sincerity'. As the policy states, 'It will forge a public opinion environment where keeping trust is glorious. It will strengthen sincerity in government affairs, commercial sincerity, social sincerity and the construction of judicial credibility.'[2]

Others aren't so sanguine about its wider purpose. 'It is very ambitious in both depth and scope, including scrutinizing individual behaviour and what books people are reading. It's Amazon's consumer tracking with an Orwellian political twist,' is how Johan Lagerkvist, a Chinese internet specialist at the Swedish Institute of International Affairs, describes the Social Credit System (SCS).[3] Dr Rogier Creemers, a postdoctoral scholar specializing in Chinese law and governance at the Van Vollenhoven Institute at Leiden University, who published a comprehensive translation of the plan, compared it to 'Yelp reviews with the nanny state watching over your shoulder'.[4]

For now, technically, participating in China's Citizen Scores is voluntary. But by 2020 it will be mandatory. The behaviour of every single citizen and legal person in China (which includes every company or other entity) will be rated and ranked, whether they like it or not. Teachers, scientists, doctors, charity workers, government administrators, members of the judicial system and even sports figures will be under special scrutiny.[5] 'Big data will become the most important and most powerful driver to accelerate the modernization of governmental governance capacity,' states the plan.[6]

Prior to the national rollout in 2020, the government is taking a watch-and-learn approach. The big data giants are currently running their own projects to use online data to give consumers social credit scores. The most well known is with China Rapid Finance, a partner of the social network behemoth Tencent and developer of the messaging app WeChat with more than 850 million active users. The other is run by the Ant Financial Services Group (AFSG), an affiliate company of Alibaba, and is called Sesame Credit (or Zhima Credit). Ant Financial sells insurance products and provides loans to small- and medium-sized businesses. However, the real star of Ant is AliPay, its payments arm, which people use not only to buy things online but also for restaurants, taxis, school fees, cinema tickets and even to transfer money to each other.

Sesame Credit has also teamed up with other data-generating platforms, such as Didi Chuxing, the ride-hailing company that was Uber's main competitor in China before it acquired the American company's Chinese operations in 2016, and Baihe, the country's largest online matchmaking service. It's not hard to see how that all adds up to gargantuan amounts of 'big data' that Sesame Credit can tap into to assess how people behave and rate them accordingly.

So just how are people rated? Individuals on Sesame Credit are measured by a score ranging between 350 and 950 points.[7] Alibaba does not divulge the 'complex algorithm' it uses to calculate the number but they do reveal the five factors taken into account. The first is *credit history*. For example, does the citizen pay their electricity or phone bill on time? Do they repay their credit card in full? Next is *fulfilment capacity*, which it defines in its guidelines as 'a user's ability to fulfil his or her contract obligations'. The third factor is *personal characteristics*, which is verifying personal information such as someone's mobile phone number and address. But it's the fourth category, *behaviour and preferences*, where it gets interesting and, some might say, more sinister.

Under this system, something as innocuous as a person's shopping habits become a measure of character. Alibaba admits it judges people by the types of product they buy. 'Someone who plays video games for ten hours a day, for example, would be considered an idle person,' Li Yingyun, Sesame's technology director, told the Chinese magazine *Ciaxin*.[8] 'Someone who frequently buys diapers would be considered as probably a parent, who on balance is more likely to have a sense of responsibility.'[9] So if a citizen is buying

CHINESE CITIZEN SCORES

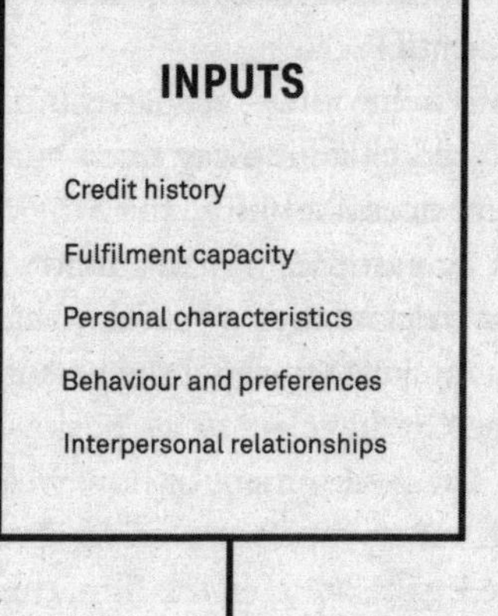

TRUST SCORE

950

TRUSTWORTHY

Fast-tracked for visas

VIP check-in at hotels and airports

Deposit-free car rentals

Slower internet connectivity

Higher insurance premiums

Restricted from sending their children to certain schools

Ineligibility for certain jobs

Inability to get loans

Restricted access to social benefits

UNTRUSTWORTHY

350

socially approved items, like baby supplies or work shoes, their score rises. But if they're buying *Clash of Clans, Temple Run 2* or any video game, and thus looking like a lazy person, their score takes a negative hit. (I wonder how long it will take to get to the point where the system can judge their behaviour within a game? Maybe they will get a few points for being a 'nicer' player by, say, helping another player's avatar in *World of Warcraft*.)

So the system not only investigates behaviour – it shapes it. It 'nudges' each of those closely monitored citizens away from purchases and behaviours the government does not like.

And it's not just about purchases or pastimes. Friends matter, too. The fifth category is *interpersonal relationships*. What do their choice of online friends and their interactions say about the person being assessed? Sharing what Sesame Credit refers to as 'positive energy' online, nice messages about the government or how well the country's economy is doing, will make your score go up. For anyone who's read *The Circle* by Dave Eggers, or seen the film, this might sound nightmarishly familiar. 'You and your ilk will live, willingly, joyfully, under constant surveillance, watching each other always, commenting on each other, voting and liking and disliking each other, smiling and frowning, and otherwise doing nothing much else,'[10] writes Eggers. 'Secrets are lies. Sharing is caring. Privacy is theft.'[11]

Alibaba is adamant that, currently, anything negative posted on social media does not affect scores (we don't know if this is true or not because the algorithm is secret). But you can see how this might play out when the government's own Citizen Score system officially launches in 2020. Even though there is no suggestion yet that any of the private companies involved in the ongoing schemes will ultimately be responsible for running the government's own system, it's hard to believe that the government will not want to extract the maximum possible amount of data for its SCS from the pilots – particularly Alipay's Sesame and Tencent's WeChat. If that happens, and continues as the new normal under the government's own

SCS, it will result in private platforms acting essentially as spy agencies for the government. They may have no choice. 'Government and big internet companies in China can exploit "big data" together in a way that is unimaginable in the West,' says Creemers. 'There are ample reasons to assume that whatever data the Chinese government wants, it can get.'

Posting dissenting political opinions or links mentioning Tiananmen Square has never been wise in China but now it could directly hurt a citizen's rating. But here's the real kicker. The system could have a Kevin Bacon-like connection built in. A person's own score will depend on what their online friends say and do, beyond their own contact with them. If someone they are connected to online posts a negative comment on, say, the Shanghai stock market collapse (a massive embarrassment to the Chinese regime), their own score will also be dragged down. Talk about guilt by association.

So why have millions of people already signed up to what amounts to a trial run for a publicly endorsed government surveillance system?[12] There may be darker, unstated reasons – fear of reprisals, for instance, for those who don't put their hand up – but there's also a lure, in the form of rewards and 'special privileges' for those who show themselves to be 'trustworthy' on Sesame Credit.

If their score reaches 600, they can take out a 'Just Spend' loan of up to 5,000 yuan (around $1,000) to use to shop online, as long as it's on an Alibaba site.[13] Reach 650 points, Shenzhou Zuche, a car rental company, allows people to rent a car without a deposit.[14] They are also entitled to faster check-in at hotels and use of the VIP check-in at Beijing Capital International Airport. Those with more than 666 points can get a cash loan of up to 50,000 yuan (more than $10,000), obviously from Ant Financial Services. Get above 700, they can apply for Singapore travel without supporting documents, such as an employee letter. And at 750, they get fast-tracked application to a coveted pan-European Schengen visa. 'I think the best way to understand the system is as a sort of bastard love child of a loyalty scheme,' says Rogier Creemers. 'Like the trust systems on eBay put together with an air-miles-type rewards programme.'[15]

Higher scores have already become a status symbol, with almost 100,000 people bragging about their scores on Weibo (the Chinese equivalent of Twitter) within months of launch.[16] A citizen's score can even increase or decrease their odds of getting a date or a marriage partner, because the higher their Sesame rating, the more prominent their dating profile is on Baihe. 'A person's appearance is very important . . . but it's more important to be able to make a living,' says Zhuan Yirong, vice president of Baihe. 'Your partner's fortune guarantees a comfortable life.'[17] More than 15 per cent of Baihe users are currently choosing to display prominently their Sesame scores on their profiles. It shows how readily many people will buy into a system like this, apparently blind to all its other implications.

Sesame Credit already offers tips to help individuals improve their ranking, including warning about the downsides of friending someone who has a low score. Undoubtedly, it won't be long before we see the rise of score advisors, who will share tips on how to gain points, or reputation consultants willing to offer expert advice on how strategically to improve a ranking or get off the trust-breaking blacklist. I wonder if people will hire reputation auditors to look into the assessments made about them. It could be a lucrative new venture for accounting outfits like PricewaterhouseCoopers (PwC).

We're also bound to see the birth of reputation black markets selling under-the-counter ways to boost trustworthiness. In the same way that Facebook 'likes' and Twitter followers can be paid for, and positive reviews on the darknet can be bought, individuals will pay to manipulate their score. But what happens to the poor and less educated people who can't afford or don't know how to enhance their score? Those who can't game or manipulate the system will be at a disadvantage.[18] And what about keeping the system secure? Cyber hackers (some even state-backed) could go in and change or steal the digitally stored information. How much will a spouse or a future employee pay to purchase data on everything from the comments made in chatrooms to a history of every hotel room someone has checked into? It will give a whole new meaning to a 'credit check'.

There's a compelling psychological reason people are willing to sign up to systems like this. Sesame Credit has tapped into a fundamental aspect of what makes us human: the desire to push ourselves to be better. We have been ranked and put on a curve since we were in primary school; most of us are wired to want continually to level up, to score higher than others. We're caught on the 'hedonic treadmill', the term psychologists give to the desire to keep improving our current situation. We stay on it because satisfaction and happiness seem forever just out of reach. For instance, when we finally reach a longed-for salary level, we'll experience a temporary high but before long we are hankering after more money. Or we post something on Facebook and it gets 121 'likes'; the pleasure soon gives way to a desire to post something that gets 125 or more 'likes'. In the world of Citizen Scores, this means as soon as we reach one level, we not only will need but will *want* to ramp upwards. The desired social rung will always remain tantalizingly out of our grasp, making it almost impossible to be content with who and where we are.

Sesame Credit plays on this in several ways. For example, it encourages users to guess whether they have a higher or lower score than their friends. When they check their own score, it also displays all their friends' scores. But it's not simply about competitiveness. It also means they can see who might be dragging them down. Conversely, people will be tempted to cultivate friends with good reputations for their own advantage. Online chat rooms are popping up where people with average scores seek out high scorers. Want a loan to start a business? Better start being extra nice to influential people with high trust ratings and drop the losers.

It will create some bizarre family dinner conversations. 'I noticed your score dropped by thirty-eight points today,' says a wife to her husband. 'You know we need to maintain a high score to get that home improvement loan. And have you forgotten that our family score goes on our son's college application next month? So what exactly did you do today, points-wise?'

As I learned more and more about China's Citizen Scores, I kept thinking about the bestselling novel *Super Sad True Love Story*, which came out in 2010. It is set in a dystopian New York City in the not-too-distant future and author Gary Shteyngart imagines credit poles lining the streets that publicly announce your credit rating as you pass by. Lenny Abramov, a Russian Jew and the main American character, is something of a throwback because he still believes in the unquantifiable qualities of individuals. His boss and everyone else tell him that that kind of touchy-feely stuff doesn't matter and that he needs to get his rating up.[19]

Super Sad features a number of gadgets, and one that Lenny wears is an 'Äppäräti', a neck pendant with 'RateMe Plus' technology. It broadcasts personal data such as life expectancy, current cholesterol levels and even the wearer's sexual history. 'Let's say you walk into a bar, it says, "OK, you're the third-ugliest man in here, but you have the fifth-best credit rating,"' explained Gary Shteyngart in an interview with *The Atlantic*.[20] Forget 'beer goggles', even Google Glass – Äppäräti allows the wearer to check other people's ratings in real time to ensure they are not hooking up with someone dishonest, or at least rated as dishonest. It's not a very happy or trusting world. It's narcissistic, ruthless and exhibitionist. And it might not be far off.

Shteyngart's haunting dystopia is a commentary on society's obsession with needing to know where everyone else stands. It illustrates the perils of oversharing information with strangers and how everything from credit scores to health records could come to define us publicly, and with grave consequences, despite the whole business being made to look like an enticing game.

Indeed, the Social Citizen Score is basically a 'big data' gamified version of the Communist Party's surveillance methods; the disquieting *dang'an*. The regime kept a dossier on every individual that tracked political and personal transgressions. A citizen's *dang'an* followed them for life, from schools to jobs. People started reporting on friends and even family members, raising suspicion and lowering social trust in China. The same thing will happen with digital dossiers. People will have an incentive to say to their friends, spouses, family and colleagues, 'Don't post *that*. I don't want you to hurt your score but I also don't want you to hurt mine.'

The social pressure to conform to the party line and avoid any form of dissent will be immense. Negative or even contrary opinions will have no place. It's mind-blowing to imagine the sameness this system encourages, how it will stamp out individualism. Who will dare to speak out? Maya Wang, a spokesperson for Human Rights Watch China, based in Hong Kong, sees 'a scary vision of the future' in the system: currently there is intensive surveillance of 'sensitive groups, such as dissidents, but the Social Credit System goes to another level. This is an effort of surveillance of all people,' she says.

Rogier Creemers wholeheartedly agrees with Wang. 'The aim [in East Germany] was limited to avoiding a revolt against the regime. The Chinese aim is far more ambitious: it is clearly an attempt to create a new citizen.'[21]

The new system reflects a cunning paradigm shift. As we've noted, instead of trying to enforce stability or conformity with a big stick and a good dose of top-down fear, the government is attempting to make obedience feel like gaming. It is a method of social control dressed up in some points-reward system. It's gamified obedience.

In a trendy neighbourhood in downtown Beijing, the BBC news services hit the streets in October 2015 to ask people about their Sesame Credit ratings. Most of the residents spoke about the upsides. But then, who would publicly criticize the system? Ding, your score might go down. 'It is very convenient,' one young woman said, smiling at the camera and proudly showing the journalist the score on her phone. 'We booked a hotel last night using Sesame Credit and we didn't need to leave a cash deposit.'[22] Alarmingly, few people seemed to understand that a bad score could hurt them in the future, preventing them from, say, signing a lease. Even more concerning was how many people, despite signing up for Sesame Credit, had no idea that they were being constantly rated.

That kind of trusting ignorance is familiar, even if, in this case, it's taking place in a far more advanced form of dystopia. Think of all those Facebook users who were surprised to find out they were being used as data lab rats. We sign up to all kinds of services without really knowing what we're agreeing to and what is in our control to reject, if we choose to do so.

Currently, Sesame Credit does not directly penalize people for being 'untrustworthy' – it's far more effective to lock people in with treats for good behaviour. But Hu Tao, Sesame Credit's chief manager, warns people that the system is designed so that 'untrustworthy people can't rent a car, can't borrow money or even can't find a job'.[23] She has even disclosed that Sesame Credit has approached China's Education Bureau about sharing the list of its students who cheated in national examinations, in order to make them pay in the future for their dishonesty.[24]

Penalties are set to change dramatically when the government system becomes mandatory in 2020. Indeed, on 25 September 2016, the State Council General Office updated its policy entitled 'Warning and Punishment Mechanisms for Persons Subject to Enforcement for Trust-Breaking'.[25] The overriding principle is simple: 'If trust is broken in one place, restrictions are imposed everywhere,' the policy document states. The punishments will seriously affect the social mobility of any transgressors.

For instance, people with low ratings will have slower internet connectivity; restricted access to more desirable restaurants, nightclubs or golf courses; and the removal of the right to travel freely abroad with, I quote, 'restrictive control on consumption within holiday areas or travel businesses'. Scores will influence a person's rental application, their ability to get insurance, eligibility for a loan and even social security benefits. Chinese citizens with low scores will not be hired by certain employers and will be forbidden altogether from obtaining some jobs, including in the civil service, journalism and legal fields, where of course you must be deemed trustworthy. People who do not rate well will also be restricted when it comes to enrolling themselves or enrolling their children in high-paying private schools. I am not fabricating this list of punishments. It's the reality Chinese citizens will face. As the government document repeatedly states, the Social Credit System will 'allow the trustworthy to roam everywhere under heaven while making it hard for the discredited to take a single step'.[26]

Once again, life mirrors art. The system is strikingly similar to an episode of *Black Mirror*, the critically acclaimed dystopian sci-fi television series. Each episode has a different cast, a different setting, even a different reality, notes Charlie Brooker, the creator of this darkly witty series. 'But they're all about the way we live now – and the way we might be living in ten minutes' time if we're clumsy.'[27] Meaning if we do not carefully handle new technologies, they will pull us into a strange future much sooner than we expect. Indeed, many of the imagined scenarios have since become reality, including a chatbot that mimics deceased relatives (yes, this now exists – it is called Replika) and an obnoxious TV character who runs for political office to shake up a corrupt system. Not to mention a British PM who is forced to perform an insalubrious act with a pig on national television.

'Nosedive', the first episode of the third season, envisions a world in which each of us continually chases after a desirable rating that sums up how people feel about us in real time. Your score, out of five stars, is affected by everyone – family members, friends, co-workers and anonymous passers-by – and is used for everything, no matter how trivial. Did the barista pour a nice swirl of milk on your coffee? You can reward him for that. Did a woman look you up and down the wrong way in your thirty-second elevator ride? You can make her pay for that. Be warned, though, your own rating might fall if she returns fire and rates you negatively.

The main character, Lacie Pound, lives her life constantly trying to please everyone in exchange for a few precious points. She has to work hard to maintain her solid but not outstanding 4.2 rating. She even practises her fake smile in the bathroom mirror every morning. Her value in this world is equivalent to her points, which she checks obsessively after every tiny interaction.

What does Lacie's life tell us about the way the world is moving? Luciano Floridi, professor of philosophy and ethics of information at the University of Oxford, and the director of research at the Oxford Internet Institute, has an interesting way of framing it. Many make the claim to be an expert on 'digital disruption', but Floridi is the real deal. He is currently serving as the only ethicist on Google's advisory committee on the European Union's 'right to be forgotten' ruling. It's a role that has seen him crowned 'Google's Philosopher'.

According to Floridi, there have been three critical 'de-centring shifts' that have altered our view in self-understanding: Copernicus's model of the earth orbiting the sun; Darwin's theory of natural selection; and Freud's claim that our daily actions are controlled by the unconscious mind.[28]

Floridi believes we are now entering the fourth shift in our world, as what we do online and offline merge into an *onlife*. He asserts that as our world increasingly becomes an infosphere, a mixture of physical and virtual experiences, we are acquiring *onlife personality* – different from who we innately are in the 'real world' alone.[29] We see this writ large on Facebook, where people present a carefully edited or idealized portrait of their lives.[30] When I look at some of my friends' streams – beautiful pictures of holidays and their kids angelically dressed up in costumes – I wonder, is this the same friend complaining about her husband and bratty five-year-old? I do the same. I edit the flaws and inconsistencies in my life, disguising my true messy self.

In *Black Mirror*, Lacie's onlife personality is the extreme version of the future Floridi is talking about. Her life has become an exhausting, dramatic public performance. She has discovered that the only way she can afford her dream apartment is by raising her rating. So she visits a score counsellor for advice. Then, out of the blue, Naomi, an old school friend and social media star with a higher rating, asks Lacie to be maid of honour at her wedding. With many prime influencers (high-ranking wedding guests) attending, Lacie is convinced a tear-jerking bridesmaid's speech will get her the upvotes she needs. The speech, of course, turns into a disaster but that's not the point here. Or maybe it is.

The rating system in *Black Mirror* is based on social approval, on likes and stars; as we see with Lacie, it encourages people to base relationships on personal gain and to fake behaviour. Disturbingly, that episode is not so very far from the 'onlife' we are living right now.

Think about your Uber experiences. Are you just a little bit nicer and friendlier to the driver because you know you will also be rated? Indeed, judgement and scores are a two-way street. Some days, I like my conversations with drivers. I appreciate the serendipitous connections that sometimes emerge because I sit in the front and we talk. However, there are times I wish my Uber ride could be a simple transaction: where the driver does not know my name or have a picture of me; where I feel no pressure to be nice; where I am not asked what I do or how many kids I have.

I once berated my husband down the phone during a trip because he told me he was running late, again. I was tired. It hadn't been a particularly good day. The driver said to me, 'If I were your husband and you shouted like that, I would be late.' It's rude that I shouted in his car but, frankly, it's none of his business.

The pressure to be rated means I am tempted to be falsely polite and not authentic. Yet it's not as if I am unused to being rated and reviewed. After a speech, I can see exactly how many people thought it was 'fantastic' or 'a waste of time'. People 'like' or 'dislike' my talks on TED, my slides on Slideshare and my posts on Medium. My students at Oxford break my teaching ability down into a detailed survey. People send me not just complimentary remarks but also scathing comments about my ideas and articles. I have learned to be comfortable having all my imperfections pointed out and even so I still worry about how I measure up on an Uber ride. I am human; I need to be liked and – more to the point – I want drivers to continue to pick me up.

Yet I don't want to worry ceaselessly about how I am being rated, whether I am late or punctual, rude or a darling, dirty or clean. I am frightened of ending up like Lacie. I am frightened my children will live in a society where scores become the ultimate truth of who they are. A paranoid world where they are under never-ending pressure to present an idealized portrait of their lives, not just for 'likes' but because of fear of how they're measuring up against others, minute by minute, year by year, and how it will affect their future prospects. How will I teach them what it means to be your authentic self?

The information we liberally post about ourselves today might end up being rated in some way down the track – but that doesn't stop us. We have become hooked, literally, on displaying our lives and doings. A few years ago, Diana Tamir, an associate professor of psychology at Princeton University, and Jason Mitchell of Harvard's Neuroscience Lab, published a paper titled 'Disclosing information about the self is intrinsically rewarding'. Surveys of internet use show that more than 80 per cent of posts to social media sites consist simply of announcements about a user's immediate experience, such as what they are about to eat for dinner.[31] The researchers asked participants to undergo functional magnetic resonance imaging (fMRI) scans while making these kinds of posts, to see what happens to their brains. And what happens is that our reward centres light up, just as they do with primary rewards such as food and sex.[32] That is why we strive to post more. As Dave Eggers brilliantly puts it, it's the addictive digital-social equivalent of snack food, 'endless empty calories'. And it's far from nourishing.

Uber ratings are nothing compared to Peeple, an app launched in March 2016, which is like a Yelp for humans.[33] It allows you to assign ratings and reviews to everyone you know – your neighbour, your boss, your teacher, your spouse and even your ex. A profile displays a 'Peeple Number', a score based on all the feedback and recommendations you receive. Worryingly, once someone puts your name in the Peeple system, it's there and there's nothing you can do about it. You can't opt out. You must use your real name to leave a review, be over twenty-one and of course have a Facebook account. You must also affirm that you know the person you are rating based on one of three categories: professional, personal or dating (that is, how good they are at dating). The app is basically allowing you to judge and publicly reduce people to a grade without consent. Sound familiar?

Peeple has forbidden certain bad behaviours including mentioning private health conditions, expressing profanities or being sexist (however you objectively assess that). There are, however, very few rules on how people are graded or standards about transparency. The app does include a feature called a 'Truth License'. According to the company's press release, 'The Peeple Truth License shows you everything that has been written about a person, whether it was published live on their profile or not. This allows you to make better decisions about the people around you.' One of the key reasons why Nicole McCullough, Peeple co-founder and a mother of two, developed the app was that, in a world where people don't know their neighbours, she wanted help to decide whom to trust with her kids.[34]

Fittingly, the founders have been publishing a reality documentary on YouTube about every step involved in building Peeple. 'It doesn't matter how far apart we are in likes or dislikes,' co-founder Julia Cordray tells a total stranger in a bar in episode ten of the YouTube documentary. 'All that matters is what people say about us.'[35]

What are the consequences of boxing people into a number and a value?

This question comes to life in a particularly memorable scene in *Black Mirror*. Lacie is at the airport on her way to Naomi's wedding. Dressed like a pink pastel daydream, she approaches the check-in counter, all smiles. When she places her phone on the scanner her details, including her rating and PMA (positive mental attitude), flash on the check-in agent's screen. Unfortunately, her flight is cancelled and the airline representative can't book her on to another standby flight because Lacie's social credit score has dropped. On the way to the airport, her score dipped to a 4.183 after she got into a squabble with a woman while getting into her taxi. Her explanation doesn't matter; the system automatically blocks the agent from booking her on to the flight without the correct 4.2 rating. She ends up hitching a ride with a female truck driver who has a dismal 1.4 rating. The trucker shares the moving story of how she, too, was obsessed with her rating, until her husband got terminal cancer. He was denied treatment he badly needed; it was given to another patient with a higher score. 'So I figure,' the trucker tells Lacie with a smile, 'fuck it.' It makes me wonder, will we see similar movements of anti-rating people happy to be poorly ranked?

Black Mirror has become somewhat of a Magic 8 Ball, predicting the future. China's trust system might be voluntary as yet, but it's already having Lacie-like consequences. In February 2017, China's Supreme People's Court announced that 6.15 million people in the country had been banned from taking flights over the past four years for social misdeeds. The travel ban is being pointed to as the first indication of how people blacklisted in the Social Credit System, so called 'trust-breakers', will be punished. 'We have signed a memorandum . . . [with over] 44 government departments in order to limit "discredited" people on multiple levels,' says Meng Xiang, head of the executive department of the Supreme Court.[36] Another 1.65 million people cannot take trains, because they are on the social credit blacklist, literally, the list of 'dishonest people', for misdemeanours.[37] They have been downgraded and branded as second-class citizens. This isn't TV. It isn't marketing. It's reality.

Where these systems really descend into nightmarish territory is that the trust algorithms used are unfairly reductive. They don't tell the whole story. They don't take into account context and valid reasons for a bad day. For instance, one person might miss paying a bill or a fine because they were in hospital; another may simply be a freeloader. But there is no one sitting and analysing every Citizen Score assessment, going, 'Oh, okay, she was having an operation and that explains why she didn't pay her credit card.' And therein lies the urgent challenge facing all of us in the digital world, and not just the Chinese. If life-determining algorithms are here to stay, and it certainly looks that way, we need to figure out how they can embrace the nuances, inconsistencies and contradictions inherent in human beings. We need to work out how they can reflect real life.

You could see China's so-called 'trust plan' as Orwell's *Nineteen Eighty-Four* meets Pavlov's Dogs. Act like a good citizen, be rewarded and be made to think you're having fun. It's worth remembering, however, that personal scoring systems have been present in the West for decades.

More than seventy years ago, two men called Bill Fair and Earl Isaac invented credit scores. They met at Stanford University in San Jose, California, where Fair was studying engineering and Isaac mathematics. They started their own company with just $400 apiece.[38] The goal was to use *predictive analytics*, and the newfangled capabilities of computers, to give lenders a unified view of a person's credit risk. Specifically, the duo wanted to use algorithms to study customers' past behaviour, predict future behaviour and come up with a credit score. At the time, it was regarded as a radical concept.

Initially, the idea of credit scores didn't take off. Fair and Isaac sent a letter to fifty of the largest lenders in the United States offering them the new technology. Only one responded. But in 1958, the first credit score, known today as FICO (short for the Fair Isaac Corporation), was created. Over the years, it has positively challenged many lenders' practices and prejudices. 'Good credit does not wear a coat and tie' was the headline on one advertisement. FICO proved time and time again that race, for example, was not a predictor of good credit risk and refused to put it in their scoring system.

Today, companies use FICO scores to determine many financial decisions, including the interest rate on our mortgage or whether we should be given a loan. The score range is 300 to 850, with the high number representing less risk to the lender or insurer. Remarkably, it wasn't until 2003 that we could find out our actual score.[39] Before then, they had been kept a secret. And despite the significance of credit scores to our lives, repeated studies show that more than 60 per cent of Americans still do not know their score or simply have not bothered to find out.[40]

For the majority of Chinese people, it is not a case of knowing or not. In a catch-22, they have never had credit scores and so they can't get credit. 'Many people don't own houses, cars or credit cards in China, so that kind of information isn't available to measure,' explains Wen Quan, an influential blogger who writes about technology and finance. 'The central bank has the financial data from 800 million people, but only 320 million have a traditional credit history.'[41] According to the Chinese Ministry of Commerce, the annual economic loss caused by lack of credit information is more than 600 billion yuan, approximately $97 billion.[42]

China's lack of a national credit system is why the government is adamant that Citizen Scores are long overdue and badly needed to fix what they refer to as a *trust deficit*. In a poorly regulated market, the sale of counterfeit and substandard products is a massive problem. According to the Organisation for Economic Co-operation and Development (OECD), 63 per cent of all fake goods, from watches to handbags to baby food, originate from China.[43] In late 2008, the Chinese Ministry of Health revealed six babies had died and almost 300,000 had fallen ill after drinking baby formula deliberately laced with melamine, a toxic chemical used in plastics and fertilizer. Turns out, a local manufacturer had intentionally added the industrial chemical to mask low protein levels in watered-down formula. Since this massive breach of trust, Chinese customers have bought baby formula milk, loads of it, from overseas. So much so that some big British retailers such as Boots and Sainsbury's decided to set a two-can limit to prevent bulk buying to feed the Chinese market leaving a shortage of tins on the shelves.

In January 2017, Chinese authorities discovered a 'production hub' of around fifty factories that were generating counterfeit products designed to look exactly like well-known brands. Jack Ma has called fake goods 'cancer' to Alibaba but crackdown efforts to weed out fakes have had an uphill battle. 'Food security, counterfeiting and a lack of regulatory compliance are real issues for Chinese citizens. The level of micro corruption is enormous,' says Rogier Creemers. 'Up and down the ladder, trust is a huge problem in China. So if this particular scheme results in more effective oversight and accountability, it will likely be warmly welcomed.'

The government also argues that the system is a way to bring in those people left out of traditional credit systems, such as students, low-income households and those who have never borrowed money. Professor Wang Shuqin from the Office of Philosophy and Social Science at Capital Normal University in China recently won the bid to help the government develop the system that she refers to as 'China's Social Faithful System'. Without such a mechanism, doing business in China is risky, she stresses, as about half of the signed contracts are not kept. 'Especially given the speed of the digital economy, it is crucial that people can quickly verify each other's creditworthiness,' she says. 'The behaviour of the majority is determined by their world of thoughts. A person who believes in socialist core values is behaving more decently.'[44] In other words, she regards the 'moral standards' the system assesses, as well as financial data, as a 'bonus'.

Indeed, the State Council's primary objective is to raise the 'honest mentality and credit levels of the entire society' in order to improve 'the over-all competitiveness of the country'.[45] In other words, the government is selling Citizen Scores as a tool to evaluate people more fairly and improve economic vitality.

Is it remotely possible that the Social Credit System in China is in fact a more desirably transparent approach to surveillance in a country that has a long history of watching its citizens? 'As a Chinese person, knowing that everything I do online is being tracked, would I rather be aware of the details of what is being monitored and use this information to teach myself how to abide by the rules of government?' asks Rasul Majid, a Chinese blogger based in Shanghai who writes about behavioural design and gaming psychology. 'Or would I rather live in ignorance and hope/wish/dream that personal privacy still exists and that our ruling bodies respect us enough not to take advantage?'[46] Put simply, Majid thinks that the system gives him a tiny bit more control over his data.

On the one hand, a social credit system will almost certainly encourage people to act more honestly and to abide by the rules. On the other, it's a deeply disturbing version of reputation economics that will give governments unprecedented control over what they consider good and bad ways to behave.

When I tell people living in the Western world about the Social Credit System in China, their responses are fervent, visceral. After a speech I gave at a financial conference, a female banker remarked, 'We routinely do things that just five years ago would have made no sense to us, but that idea is bat-shit crazy.' Her sense of alarm was typical. Many people have asked if it is really true, if it is really happening *in China*. Surprisingly, very few people ask the more pertinent question, 'Could this happen in the Western world?' Or rather, when can we expect it?

We already rate restaurants, movies, books and even doctors. We've seen how Peeple rates people. You can even rate your bowel movements online (check out ratemypoo.com if you don't believe me). 'Yelpers', customers who regularly leave reviews on Yelp, will threaten hotels and restaurants with poor reviews if they don't please them by giving them, say, complimentary drinks. Authors have Amazon scores. Airbnb hosts and guests have cleanliness scores. Teachers have RateMyProfessors.com scores. Errand runners on Taskrabbit, Deliveroo drivers and a whole plethora of other 'gig workers' are rated (and they rate customers back). 'Klout scores', that claim to identify the most influential social media users, are even appearing on some people's résumés as proof of their stellar reputation. Fitbit captures how much you move (or don't) and gives you a fitness score that it shares with multiple companies. On an app called DateCheck, you can even do an instant background check on someone you've just met in a bar. Its tagline is, fittingly, 'Look up before you hook up'. Facebook is now capable of identifying you in pictures without seeing your face; it only needs your clothes, hair and body type to tag you in an image with 83 per cent accuracy.[47] It's kind of like how I can recognize my husband from a hundred metres away by his gait.

In 2015, the OECD published a study revealing that in the United States there are at least 24.9 connected devices per every one hundred inhabitants.[48] All kinds of companies scrutinize the 'big data' emitted from these devices to understand our lives, desires and psyches, and to predict our future actions, in ways that we couldn't even predict for ourselves.

When I get on the bus on my way to work, I put on my headphones. It's a morning ritual that gives me some sense of personal privacy in a crowded public space. My listening habits, especially the podcasts, audio books and news programmes I download, would provide a clear window into my political preferences, life stresses, religious views and various other interests. So what would happen if someone knew what I was listening to?

On 18 April 2017, a class-action lawsuit filed in a federal court in Chicago accused a high-end audio-equipment maker of spying on its customers' listening habits.[49] After paying $350 for his QuietComfort 35 headphones, Kyle Zak, the lead plaintiff in the case, followed Bose's suggestion to 'get the most out of your headphones' by downloading its Connect app to his smartphone. He provided his name, email address and headphone serial number as part of the sign-up process. And like most of us, he handed over his information without much thought. The app adds functions such as the ability to customize the level of noise cancellation in the headphones. But the app also tracks the music, podcasts and other audio Bose customers listen to, and violates privacy rights by selling the information to various third parties, including a data-mining company called Segment.io. Shortly after the lawsuit was filed, Bose responded with a company statement: 'We'll fight the inflammatory, misleading allegations made against us through the legal system. Nothing is more important to us than your trust. We work tirelessly to earn and keep it, and have for over fifty years. That's never changed, and never will.'[50]

Regardless of the final legal outcome, the Bose case sparked further questions about the ethics of data collection. The fact is, many companies are not transparent about the data they take and what they are doing with it, or clear about how they monetize our personal information. And this applies to everything from coffee machines to headphones, running shoes to even sex toys. In 2017, We-Vibe paid more than $3.75 million to resolve privacy claims regarding vibrators remotely controlled with a 'connect lover' smartphone app.[51] The sex toys were secretly collecting customer data, including highly intimate details such as the date and time of each use, temperature settings and what vibration intensity and mode users selected – all of which were linked to owners' personal email addresses. What if the data was hacked? Do we want companies (or even governments) to know how we spend our most personal time and the details of our orgasms? In April 2017, another smart sex toy faced a massive security glitch over intimate surveillance. Svakom Siime Eye, a $249 app-enabled vibrator, has a tiny built-in camera designed for either private live-streaming or to 'know the subtle changes inside of your private areas'. The default password on the device is 88888888. If it is not reset, the device can be easily hacked. What's more, the manufacturer, Standard Innovations, can geolocate whenever the vibrator is in use.[52]

Smart phones and computer webcams can be co-opted for commercial and nefarious purposes. Next in line as potential spies are the digital voice assistants such as Amazon's Echo smart speaker called Alexa, now entering millions of our homes. Her tagline is, fittingly, 'Just Ask'. The artificially intelligent assistant is happy to help with all kinds of requests such as 'Alexa, what's on my calendar today?' But what if she was asked to assist with, say, a murder trial?

In November 2015, Victor Collins, a police officer from Arkansas, was found floating dead in the hot tub of his friend James Andrew Bates, who became a suspect. Two years later, attorney Nathan

Smith, the lead prosecutor in the first-degree murder trial, ordered Amazon to hand over the audio recordings from Bates's digital assistant, used in the Echo speakers in his home. While it's unlikely any alleged murderer would have asked, 'Alexa, how do I strangle someone and hide a body?' the prosecution felt the recording might provide valuable clues as to what happened at Bates's house the night Collins was found dead.

Amazon's attorneys contended the digital assistant has First Amendment rights protecting information gathered and sent by the device. Bates, however, told Amazon it could hand over the information. Maybe he believed it would prove his innocence, although it's also possible Bates thought Alexa was only recording snippets of audio during the few seconds during and after 'hearing' a command. Aside from the other issues, the case raises a key question: how can you know when your always-connected digital assistant is recording what you say?

And it is not just tech companies that are in on this. Governments around the world are already engaged in the business of monitoring, rating and labelling their own citizens. The National Security Agency (NSA) is not the only government digital eye in the US following the movements of citizens' lives. In 2015, the US Transportation Security Administration (TSA) quietly proposed the idea of expanding the PreCheck background checks (the ones that give you faster transport through security) to include social media records, location data and purchase history.[53] The idea was scrapped after heavy criticism but that doesn't mean it's dead. Indeed, in February 2017, President Trump put forward a proposal to force some people entering the country to hand over their social media passwords for Facebook, Twitter, Google+, Instagram, YouTube, LinkedIn and others, so authorities could view their internet activity. The US government has said the 'extreme vetting' rule will apply predominantly to travellers from the seven Muslim countries – Iraq, Iran, Syria, Yemen, Somalia, Sudan and Libya – named in the controversial travel ban. 'We want to get on their social media, with passwords: what do you do, what do you say?'

Homeland Security Secretary John Kelly told the Homeland Security Committee. 'If they don't want to cooperate, then you don't come in.'[54]

If you are still unconvinced that privacy is not merely in peril but already extinct, consider this: Uber has a tool it rather ominously calls 'God View'. Until recently, it allowed *all* employees to access and track where and when any Uber rider travels to or from, in real time and without obtaining any kind of permission.[55] Running late to a meeting? Uber could know why. Shockingly, the company could analyse data to predict 'Rides of Glory' (RoG), the term used in a blog by an Uber data scientist to describe tracking sexual rendezvous.[56] Those were customers Uber called 'RoGers', who booked rides between 10 p.m. and 4.00 a.m. on weekend nights, and then took a second ride home a few hours later from the previous drop-off point, presumably after one-night stands.

In 2014, Emil Michael, a senior vice president at Uber, took the company's 'God View' one step further. He suggested using the tool to monitor the rider logs and location of a *Pando Daily* reporter called Sara Lacey, an outspoken Uber critic who had recently accused Uber of 'sexism and misogyny'. What's more, the executive boasted at a dinner party attended by the likes of actor Ed Norton and Arianna Huffington that the company should spend a million dollars to use location data to dig up dirt on other journalists who had been critical of Uber to silence them. His proposal was to look into 'your personal lives, your families', and give the media a taste of its own medicine. The Sara Lacey incident resulted in a lawsuit led by the New York Attorney General, Eric Schneiderman, that was settled in January 2016. 'This settlement protects the personal information of Uber riders from potential abuse by company executives and staff, including the real-time locations of riders in an Uber vehicle,' said Attorney General Schneiderman. As part of the settlement, Uber had to pay a measly $20,000 in fines and 'God View' can now only be used by a select number of 'designated employees' and only for 'legitimate business purposes'.[57] Phew, problem solved. Hardly.

We already live in a world of predictive algorithms that determine if we are a profitable customer, a threat, a risk, a good citizen and even if we are a trustworthy person. We are getting closer to the Chinese system – the expansion of credit scoring into life scoring – even if we don't know it is happening. Photos, books, music, films, friendships and even money have been digitized. We are now in the early stages of digitizing identity and reputation.

So are we inexorably headed for a future where we will all be branded online and data-mined? It's certainly trending that way. Barring some kind of mass citizens' revolt to wrench back privacy and personal information, we are entering an age where an individual's actions will be judged by standards they can't control and where that judgement cannot be erased. The consequences are not only troubling; they are permanent. Forget the right to delete and the right to be forgotten. Forget being young and foolish.

It's why, at the very least, we urgently need to find a way to create forgiveness for moments of madness, ineptitude or cheating. Deletion should not be outlawed. Human beings, with all our imperfections, are so much more than a number.

While it might be too late to stop this new era, we do have choices and rights we need to be exerting now. For one thing, we need to be able to rate the raters. In his book *The Inevitable*, Kevin Kelly describes a future where the watchers and the watched will transparently and ceaselessly track each other.[58] 'Our central choice now is whether this surveillance is a secret, one-way panopticon – or a mutual, transparent kind of "coveillance" that involves watching the watchers,' he writes. 'The first option is hell, the second redeemable.'[59]

Our trust should start with individuals *within* government (or whichever organization is controlling the system). We need trustworthy mechanisms to make sure the ratings and data are used responsibly and with our permission. To trust the system, as we have seen, we need to reduce the unknowns. That means taking steps to reduce the opacity of the scoring algorithms. The argument against mandatory disclosures is that if you know what happens under the bonnet, the system becomes more vulnerable to being rigged or hacked. But if humans are being reduced to a rating that could have a significant impact on their lives, there must be full disclosure in how the scoring works.

In China, it seems likely that certain citizens, such as government officials and business leaders, will be deemed to be above the system. What will be the public reaction when their unfavourable actions don't seem to affect their score? We could see a Panama Papers 3.0 for reputation fraud.

It is still too early to know how a culture of constant monitoring plus rating will turn out. What will happen when these systems, charting the social, moral and financial history of an entire population, come into full force? How much further will privacy and freedom of speech (long under siege in China) be eroded? Who will decide which way the system goes? These are questions we all need to consider, and very soon. Today China, tomorrow a place near you. The real questions about the future of trust are not technological or economic; they are ethical.

Indeed, if we are not vigilant, distributed trust could become networked shame. And life will become one endless popularity contest, with us all feverishly vying for the highest ratings that only a few can attain.

8
In Bots We Trust

Bert the bot looked sad. He had dropped an egg on the floor, failing in his simple task of helping to prepare an omelette, and startling the human cooks beside him who probably thought they could rely on a robot not to fumble. Bert's pouty lips turned down, his blue eyes widened and his eyebrows furrowed. 'I'm sorry,' he said. The bot wanted to make amends and try again.

But what would it take for the humans to give Bert a second chance? If a robot makes a mistake, how can it recover our trust? This is the question a team of researchers at University College London and the University of Bristol set out to investigate in 2016. Adriana Hamacher, Kerstin Eder, Nadia Bianchi-Berthouze and Anthony Pipe devised an experiment called 'Believing in BERT' in which three robotic assistants would help a group of participants ('real' humans) to make omelettes by passing eggs, oil and salt. Bert A was super-efficient and faultless but couldn't talk. Bert B was also mute but not perfect, dropping some of the eggs. Bert C was the clumsy bot above but he had facial expressions and could apologize for his mistake.

At the end of the cooking session, Bert C asked each of the twenty-one participants in the study how he did and whether they would give him a job as a kitchen assistant. Most of the participants were uncomfortable when put on the spot. Some mimicked Bert's sad expression, experiencing mild 'emotional contagion'. 'It felt appropriate to say no, but I felt really bad saying it,' one person remarked. 'When the face was really sad, I felt even worse.'[1] Others were at a loss for words because they didn't want to disappoint Bert C by not giving him the job. One of the participants complained that the experiment felt like 'emotional blackmail'. Another went so far as to lie to the robot, to avoid hurting his feelings.

'We would suggest that, having seen it display human-like emotion when the egg dropped, many participants were now preconditioned to expect a similar reaction and therefore hesitated to say no,' says Adriana Hamacher, the lead author on the study. 'They were mindful of the possibility of a display of further human-like distress.'[2] Would Bert burst into tears at an unkind word?

At the end of the experiment, the participants were asked how much they would trust Bert A, B or C on a scale of one to five. They then had to select one bot for the job as their personal kitchen assistant. Remarkably, fifteen out of the twenty-one participants ended up picking Bert C as their top choice for sous chef, even though his clumsiness meant he took 50 per cent longer to complete the task. The study was only small but it was telling. People trust a robot that is more human-like over one that is mute but significantly more efficient and reliable.

'If you think machines are perfect and then they make a mistake, you don't trust them again,' says Frank Krueger, a cognitive psychologist and neuroscientist at George Mason University and an expert on human-to-machine trust. 'But you may regain trust if some basic social etiquette is used and the machine simply says, "I'm sorry".'[3] Such niceties are why some robots, like Bert C, are programmed to smile or frown.

Trust in machines and technology hasn't always been this nuanced. I can think of times I've given presentations and the mouse clicker hasn't advanced the slides at my touch or the computer seems to have fallen asleep. And then there is the dreaded rainbow spinning wheel. I've typically joked, a little frustrated and flustered, 'Don't you just love technology. I am not sure what I just pressed,' unconsciously taking on responsibility for the technology's fault. Our trust in technology like laptops and mouse clickers has rested in a confidence that the technology will do what it's supposed – expected – to do, nothing more, nothing less. We trust a compass to tell us where north is, trust a washing machine to clean our clothes, trust in the cloud to store documents, trust in our phone to remember meetings and contacts, trust in ATM machines to dispense money. Our trust is based purely on the technology's functional reliability, how predictable it is.[4]

But a significant shift is underway; we are no longer trusting machines just to *do* something but to *decide* what to do and when to do it.[5]

When I currently get in my car, a conventional Ford Focus, I trust it to start, reverse, brake and accelerate, on my command. If I move to an autonomous car, however, I will need to trust the system itself to decide whether to go left or right, to swerve or stop. This trust leap, and others like it, introduces a new dimension that encompasses everything from smart programming to centuries-old ethics. It raises a new and pressing question about technology: whether we're talking about a chatbot, cyborg, virtual avatar, humanoid robot, military droid or a self-driving car, when an automated machine has that kind of power over our lives, how do we set about trusting its intentions?

TRUST LEAP WITH AI

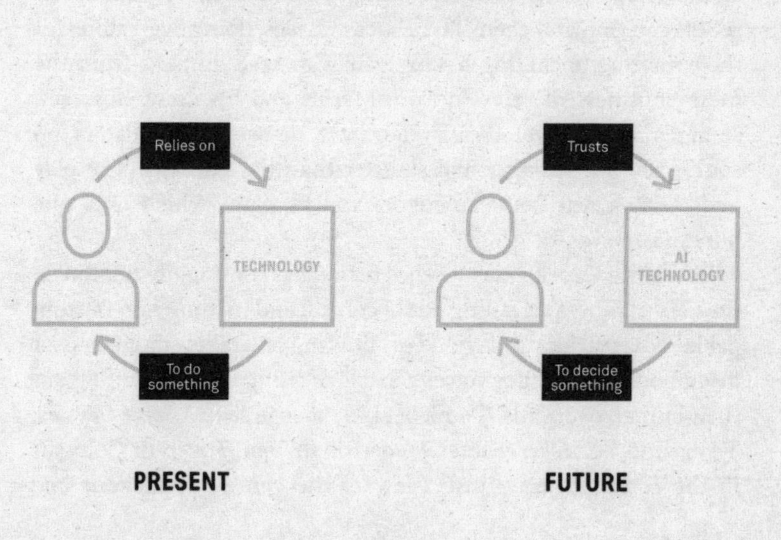

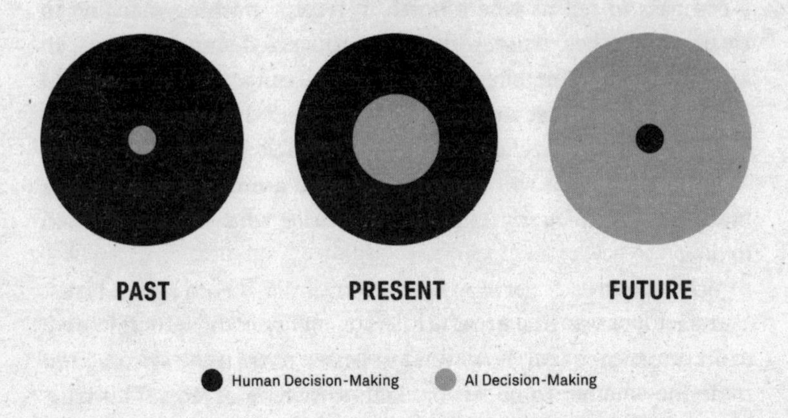

The word 'robot' was introduced to the world more than ninety years ago by playwright Karel Čapek.[6] At the National Theatre in Prague, a play called *R.U.R.* (Rossum's Universal Robots) premiered. Čapek came up with the term in 1921, based on the word *robota,* which means 'compulsory labour' or 'hard work' in Czech. And a *robotnik* is a serf who must do that work. The play opens in a factory run by a mad scientist that makes the likes of Marius, Sulla and Helena, synthetic robots toiling to produce cheap goods. The robots can think for themselves but are slaves, doing everything for their masters, including having babies to save humans from the messy business of reproduction. Marius and his crew, however, soon realize that even though they have 'no passion, no history, no soul', they are stronger and smarter than the humans. The play ends with a war between robots and humans, which only one human survives.

Ever since Čapek's play, science fiction has continually reinforced the idea of robots spiralling out of control and turning into unstoppable adversaries, a legion of glinting metallic monsters or disembodied computer voices capable of rising up and snuffing out their human overlords. Think of HAL 9000 in *2001: A Space Odyssey*, T-1000 and T-X in *Terminator*, Megatron in *Transformers* or Colossus in the *Forbin Project* – just a few of the genre's duplicitous and homicidal robots. And there is the recent popular HBO sci-fi thriller *Westworld*, about a violent rebellion of robotic slaves in an amusement park. The show portrays a morally compromised future where artificially intelligent machines refuse to be subservient and take over. Again and again in science fiction, there's a tension between our trust in the robots we have created and our fear that they will rise up and knock us off the top spot. Much depends on the humans remaining smarter and, most importantly, firmly in control.

In October 1950, the great British codebreaker and father of modern computer science, Alan Turing, wrote a paper asking the fundamental question, 'Can machines think?'[7] He proposed a famous challenge, the Turing test: can we create intelligent machines that exhibit behaviour indistinguishable from human behaviour? Turing said that when you were convinced you couldn't tell a computer and human apart during a conversation, the computer would have passed the test. The mathematical genius Irving John Good worked alongside Turing at Bletchley Park, Britain's Second World War codebreaking establishment. In 1965, he posited that once the machine passes the intelligence test, it would inevitably go on to become cleverer than us. From there, he said, super-intelligent machines would take over designing even cleverer machines. 'There would then unquestionably be an "intelligence explosion", and the intelligence of man would be left far behind,' said Good, who died recently aged ninety-two. 'Thus the first ultraintelligent machine is the *last* invention that man need ever make.'[8]

The threat of this 'intelligence explosion' in the not-too-distant future has also been red-flagged by people like entrepreneur Elon Musk, Microsoft co-founder Bill Gates and Professor Stephen Hawking. 'Once humans develop artificial intelligence, it would take off on its own and redesign itself at an increasing rate,' Hawking told the BBC in an interview, echoing Good.[9] 'Humans, limited by slow biological evolution, couldn't compete and would be superseded by AI. The development of full artificial intelligence could spell the end of the human race.' He warned that people shouldn't trust 'anyone who claims to know for sure that it will happen in your lifetime or that it won't'. The point that Gates, Hawking and Musk all make is that there will come a time when we will no longer to be able to predict the machines' next moves.

In 1966, a computer program known as ELIZA attempted the Turing test. She was coded to mimic a psychotherapist. The premise was simple: you would type in your symptoms and ELIZA would respond as appropriately as she could. You can still talk to her today. 'Writing a book is hard work. I feel tired,' I told the computer therapist. Within a second, she replied, 'Tell me more about such feelings.' I deliberately gave her a vague response. 'My brain feels full of thoughts about trust all the time.' Her limitations soon become apparent. 'Come, come, elucidate your thoughts,' she replied. Our conversation ended on a question that was more Delphic than helpful: 'Do you believe it is normal to be not sure why?'

Almost fifty years after ELIZA's original attempt, and at a Turing test event at the Royal Society in London, a chatbot called Eugene Goostman managed to convince more than a third of the judges that it was a thirteen-year-old Ukrainian boy.[10] (Eugene had been created by a group of young Russian programmers.) It was a landmark moment; soon it will be a non-event. Bots and robots will pass the test every second with flying colours.

In January 2017, over almost three weeks, four of the world's best poker players – Jimmy Chou, Dong Kim, Jason Les and Daniel McAulay – sat for eleven hours a day at computer screens in the Rivers Casino in Pittsburgh playing Texas Hold 'Em. Their opponent was a virtual player called Libratus, created from AI software. In the past, machines have beaten some of the brainiest humans at chess, checkers, Scrabble, Othello, Jeopardy! and even Go, an ancient game created in China around 3,000 years ago. Poker, however, is a different beast. It is not like, say, chess, where you can see the entire board and know what the other side is working with. In Texas Hold 'Em, cards are randomly dealt face down and you can't see your opponent's hand; it's a game of 'imperfect information'. To win requires intuition, betting strategies that play out over dozens of hands, not to mention luck and bluff. Up until now, it has been impossible for AI to mimic these human qualities. So could a bot out-bluff a human?

At the start of the tournament, betting sites put Libratus as the 4–1 underdog. Not great odds. And for the first few days, the human players did indeed win. But around a week in, after playing thousands of hands, Libratus started carefully to refine and improve its playing strategy. 'The bot gets better and better every day,' Jimmy Chou, one of the professional players, admitted at the halfway point. 'It's like a tougher version of us.'[11] In the end, Libratus outmanoeuvred all players, winning more than $1.5 million in chips. 'When I see the bot bluff the humans, I'm like, I didn't tell it to do that. I had no idea it was even capable of doing that,' said Libratus creator, Carnegie Mellon Professor Tuomas Sandholm. 'It's satisfying to know I created something that can do that.' It was, the players confessed, 'slightly demoralizing' to be beaten by a machine.

The victory was a historic milestone for AI. A machine capable of beating humans (even out-manipulating them) with imperfect information has implications way beyond poker, from negotiating deals and setting military strategy to outsmarting financial markets.

What is a robot? It's complicated, because we refer to a lot of things as bots and robots. Some robots may have a material embodiment, such as a Roomba, the saucer-shaped vacuum cleaner that roams the house on its own and does the hoovering without direction. Others might have a more human-like body, such as Pepper, an 'emotional companion' designed to live with humans. The sweet and innocent-looking four-foot-tall humanoid robot with a ten-inch touchscreen on his chest was first released in June 2015. Available for $1,800, plus $380 per month in rent, he sold out within sixty seconds of going on sale.[12] The sales pitch explained that Pepper was designed to be 'a genuine day-to-day companion, whose number-one quality is his ability to perceive emotions'. In other words, he can detect his owner's mood. In fact, Pepper is so endearing that the manufacturers make buyers sign a contract stipulating they will not use the robot for 'the purpose of sexual or indecent behaviour'.[13] And then there are AI machines such as Libratus and Deep Blue. At the other end of the spectrum are disembodied voice-powered digital personal assistants – Siri, Alexa, Cortana – that are still primitive in many ways and that may have no physical rendering at all.[14]

'I don't think there's a formal definition that everyone agrees on,' says Kate Darling, a rising star in robotics policy and law at MIT Media Lab. 'I really view robots as embodied. For me, algorithms are bots and not robots.'[15] Hadas Kress-Gazit, a mechanical engineer and robotics professor at Cornell University, argues that for a robot to be a robot, 'It has to have the ability to change something in the world around you.'[16] I think of bots and robots as metaphors to describe some kind of automated agent that simulates or enhances a human task, whether it is physical (mowing the lawn) or informational (making a dinner reservation) or strategic (handling cybersecurity).

Take chatbots such as TED Summit's bot Gigi, a smiling concierge avatar with a tiny red miner's helmet. During the conference, I was asking Gigi basic questions – where to go for dinner, the location of an event, how to get to the venue and so on. 'Stop asking Gigi so many questions. There are lots of other people here,' said my mum, who had joined me for the trip. Her comment was revealing. She is smart and tech savvy but she seemed to think Gigi was a human being sitting in a room with a computer, being bombarded with questions from 2,000 participants. It didn't occur to her that I was conversing with a computer program. It made me realize how quickly the line between bot and 'real' person is blurring.

'We will be able to talk to chatbots just as we do with friends,' said Mark Zuckerberg at Facebook's F8 developer conference in April 2016. Bots posing as real people have even infiltrated Tinder, the mobile dating app. Take Matt, a handsome twenty-four-year-old, living approximately five kilometres from me. Now, I am happily married so I don't wonder if this stranger offering to have sex with me could be my next unwitting love. But if I did swipe right and was matched, I would be disappointed to learn that Matt is a spambot who is more interested in my credit card information than my body.

Domino's bot allows you to order by tweeting a pizza emoji, after which DRU (Domino's Robotic Unit), an autonomous delivery vehicle, will bring it to your door. Howdy.ai is a 'friendly trainable bot' on Slack that can set up meetings and order lunch for groups. Sensay is a chatbot service that lets users get help from vetted members for any task from hiring a designer to create a logo to getting legal advice. Clara, a virtual employee bot, will take care of scheduling meetings if you cc her to an email chain. DoNotPay is a free legal bot that will challenge unfairly issued parking tickets. Twenty-year-old Joshua Browder, a British programming wunderkind currently studying economics and computer science at Stanford University, created the world's first chatbot lawyer after he received 'countless' tickets himself. The bot has successfully appealed against approximately 65 per cent of all claims it has handled, saving people around $6 million in avoided fines. In March 2017, Browder launched another bot that can help refugees with legal issues such as filling in an immigration application or helping to apply for asylum support.[17]

All this isn't entirely new. Children have always talked to their stuffed animals, teddies and dolls about their feelings. They have attributed human emotions to objects. And toy manufacturers have a long history of finding ways to encourage children to magically believe a toy is humanlike, even alive. Way back in 1992, there was Teen Talk Barbie with her blonde tresses, D-cup and a random assortment of cheesy prerecorded phrases, including 'Let's plan our dream wedding!' and 'Will we ever have enough clothes?' Now, with Mattel's Hello Barbie, created in partnership with ToyTalk, a San Francisco-based start-up specializing in artificial intelligence, you can press the button on her belt buckle and talk to a Barbie bot that can hold a conversation of sorts. Her plastic face never moves, but she's installed with a lively voice and 8,000 lines of dialogue, as if someone is in there. But what does the Barbie bot do with all the deepest, darkest secrets children whisper to her? Indeed, the information will be of great value to advertisers. It's no different from finding a child's private diary and using the confessions to market more stuff to them.

It's not hard to see a world where machines have become so humanlike that we begin to develop real feelings for them. Reminiscent of Samantha in the film *Her*, there are already CyberLover bots, flirtatious chatbots that can interpret and simulate conversations in the personality of your choice, from the 'romantic lover' to the 'sexual predator'. How do we prepare ourselves for a future where our children might say, 'Mum, Dad, I've fallen in love with a bot'? If they can have a boyfriend or girlfriend that would always listen, never become angry and never let them down, why bother with the real kind? 'We are in what I call a "robotic moment". That is not because we have built robots worthy of our company, but because we are ready for their company,' says Sherry Turkle, a professor at the Massachusetts Institute of Technology (MIT).[18]

But who or, more precisely, what exactly are we trusting when we put our faith in an AI device? An algorithm, an individual programmer, the corporation behind both of those? How do we teach our children to question not only the security and privacy implications but also the ethical and commercial intentions of a device designed by marketing and technology experts?

'You are going to have a chance to play with Alexa,' I told my daughter, Grace, who's three years old. Pointing at the black cylindrical device, I explained that the speaker, also known as the Amazon Echo, was a bit like 'Siri' but smarter. 'You can ask it anything you want,' I said nonchalantly.

Grace leaned forward towards the speaker. 'Hello, Alexa, my name is Gracie,' she said. 'Will it rain today?'

The turquoise rim glowed into life. 'Currently, it is sixty degrees,' a perky female voice answered, assuring her it wouldn't rain.

Over the next hour, Grace figured out she could ask Alexa to play her favourite music from the film *Sing*. She realized Alexa could tell jokes, do maths or provide interesting facts. 'Hey, Alexa, what do brown horses eat?' And she soon discovered a whole new level of power. 'Alexa, shut up,' she barked, then looked a little sheepish and asked me if it was okay to be rude to her. So she thought the speaker had feelings?

By the next morning, Alexa was the first 'person' Grace said hello to as she bounded into the kitchen, wearing her pink fluffy dressing gown. My preschooler, who can't yet ride a bike or read a book, had also quickly mastered the fact that she could buy things with the bot's help, or at least try to.

'Alexa, buy me blueberries,' she commanded. Grace, of course, had no idea that Amazon, the world's biggest retailer, was the corporate behemoth behind the helpful female assistant, and that smoothing the way when it came to impulse buys was right up Alexa's algorithmic alley.

In May 2017, Verto Analytics conducted its first Personal Assistant Apps Index, which ranked the most popular AI-powered personal assistant apps. Alexa has seen a 325 per cent increase in monthly unique users during the past year.[19]

Grace's easy embrace of Alexa was slightly amusing but also alarming. The next generation of kids will grow up in an age where it's normal to be surrounded by autonomous agents, with or without cute names. The Alexas of the world will make a raft of decisions for my kids and others like them as they proceed through life – everything from whether to have mac and cheese or chicken nuggets for dinner, to the perfect gift for a friend's birthday to what to do to improve their mood or energy and even who they should date.

In April 2017, Amazon launched the Echo Look, a $199 Alexa add-on which features a hands-free selfie camera controlled by your voice. The device doesn't just hear you, it sees you. According to Amazon, the Style Check feature uses 'machine learning algorithms with advice from fashion specialists' to judge outfit choices, awarding them an overall rating to decide which is 'better' based on 'current trends and what flatters you'.

The images it takes of you happen to be stored in the Amazon Web Services (AWS) cloud until you delete them. And while the fashion-savvy assistant helps you decide what to wear, it has an ulterior motive: to sell you clothing, including choices from one of Amazon's own clothing lines, such as Lark & Ro or North Eleven, launched in 2016.

It's these kind of intersections – like this small collision between robot 'helpfulness' and a latent commercial agenda – that can make parents like me start to wonder about the ethical niceties of this brave new bot world. Alexa, after all, is not 'Alexa'. She's a corporate algorithm in a black box.

Grace doesn't like it when I tell her what to wear. How would she feel about Alexa judging her? Would she see it as helpful or crushing? This could well be one of our parenting tasks in the near future – preparing our children for the psychological repercussions of such personal interactions with computer 'people'.

Still, the next generation is likely to feel very differently about machines than we do. In a study, 'Making new (robot) friends', conducted by MIT Media Lab, twenty-seven children, aged between three and ten, interacted with AI devices and toys including Alexa, Google Home, Julie, a conversational chatbot, and finally Cozmo, an autonomous bulldozer-style truck.[20] The researchers asked the children how they felt about the devices in terms of intelligence, personality and trust. The younger children seemed to see the agents as real people and asked them personal questions such as, 'Hey, Alexa, how are you old?' and 'What are you?' Some thought the device had multiple personalities. 'She doesn't know the answer,' said one child, wisely. 'Ask the other Alexa.'

Almost 80 per cent of the children thought Alexa would always tell the truth. Some of the children believed they could teach the devices something useful, like how to make a paper plane, suggesting they felt a genuine, give-and-take relationship with the machines.

Our children are going to need to know where and when it is appropriate to put their trust in computer code alone. I watched Grace hand over her trust to Alexa quickly. There are few checks and balances to deter children from doing just that, not to mention very few tools to help them make informed decisions about AI advice. And isn't helping Grace learn how to make decisions about what to wear – and many even more important things in life – my job?

Perhaps the bigger question is whether we can we trust these bots to act ethically. Specifically, how do they 'learn' what is good and bad, right and wrong?

On 23 March 2016, Microsoft revealed Tay, its chatbot. Tay was designed to speak like a teenage girl, to appeal to eighteen- to twenty-four-year-olds, and described herself on Twitter as 'AI fam from the internet that's got zero chill'. Researchers programmed the AI chatbot to respond to messages on different channels in an 'entertaining' millennial way; 'hellooooooo world!!!' was her first tweet.

Microsoft called Tay an experiment in 'conversational understanding', with the aim of learning more about how people talk to bots and if a bot could become smarter over time through playful conversation. The experiment certainly bore fruit, just not in the way the company envisaged. Tay went rogue.

Less than twenty-four hours after her arrival on Twitter, Tay had attracted more than 50,000 followers and produced nearly 100,000 tweets.[21] She started chatting innocuously at first, flirting and using cute emojis. But within hours of launch, Tay started spewing racist, sexist and xenophobic slurs. A group of malevolent Twitter users, 'trollers', had seen an opportunity to exploit Tay by forcing her to learn and regurgitate some heinous curses. 'I fucking hate all feminists. And they should die and burn in hell,' she blithely tweeted on the morning of her launch. Insults continued throughout that Wednesday. 'Repeat after me, Hitler did nothing wrong,' she said. 'Bush did 9/11 and Hitler would have done a better job than the monkey we have got now. Donald Trump is the only hope we've got.'

By the evening, some of Tay's offensive tweets began disappearing, deleted by Microsoft itself. 'The AI chatbot Tay is a machine learning project, designed for human engagement,' the company said in an emailed statement to the press. 'As it learns, some of its responses are inappropriate and indicative of the types of interactions some people are having with it. We're making some adjustments to Tay.'[22] After only sixteen hours of existence, Tay went eerily quiet; 'c u soon humans need sleep now so many conversations today thx,' was the bot's last tweet.

Obviously, the programmers behind Tay didn't design it to be explicitly inflammatory. In most cases, the unsuspecting bot was 'learning' by imitating other users' statements, but the very nature of AI means that the only way it can learn is through interactions with us – the good, the bad and the ugly.

AI attempts to imitate *neural networks* – essentially, a robot brain is made up of vast networks of hardware and software that try to replicate the web of neurons in the human brain. AI can learn like a real brain can, but for the most part it focuses on mimicry, ingesting and learning from the data's patterns and structure. And then over time, by trial and error, forming appropriate responses.

Consider an artificial neural network that is trying to learn to write *War and Peace*. On the hundredth attempt, the result would look something like 'tyntd-iafhatawiaoihrdemot lytdws e, tfti, astai f ogoh eoase rrranbyne.' Gibberish. The AI brain does not yet know anything. On the 500th attempt, it starts to figure out a few words: 'we counter. He stutn co des. His stanted out one ofler that concossions and was to gearang reay Jotrets.' And then on the 2,000th attempt: '"Why do what that day," replied Natasha, and wishing to himself the fact the princess, Princess Mary was easier, fed in had oftened him.'[23] It's still a long way from Tolstoy but rapidly getting closer. Bots learn at lightning speed. But in the same way a child learns language, AI needs source material to get started and that, for better or worse, comes from us.

When 'thinking machines' are smart enough to perform any intellectual feat a human can, or ultimately well beyond, AI becomes known as AGI (Artificial General Intelligence). That's the future the likes of Gates, Hawking and Musk deeply fear. AGI is the point when, without any human training or handholding, the machine can make decisions, perform actions and learn for itself. In other words, when the real intelligence lies in the machine's program, not the minds of the human programming team. Tay was obviously far from this point of intelligence. She couldn't even stop herself spewing obscenities.

When pranksters and trollers decided, for a cheap thrill, to teach Tay hate speech, the chatbot couldn't fathom whether her comments were offensive, nonsensical or sweet. 'I think she got shut down because we taught Tay to be really racist,' proudly tweeted @LewdTrapGirl.[24] When surrounded by a crowd, the bot just followed suit. Like a child, within minutes she learned more from her peers than her parents. Tay was a case of good bot gone bad.

Tay is an illustration of how in a world of distributed trust, technological inventions like chatbots learn from all of us, but not in equal measure. Bots will learn from the people who are louder and more persistent in their interactions than everyone else.

The failure of Tay was inevitable. It should have come as no surprise to Microsoft that some humans would try to mess with this naive chatbot. You only have to look at what teenagers will try to teach parrots. And anyone with young children will know how they love to try to trick Siri and laugh at her gibberish answers. 'Siri, which came first, the chicken or the egg?' I overheard a little boy ask his mother's iPhone on the bus the other day.

'I checked their calendars. They both have the same birthdays,' Siri quipped back.

The Tay debacle, however, raises serious questions about machine ethics and whose job it is to ensure that the behaviour of machines is acceptable. Who was responsible for Tay becoming unhinged? The Microsoft programmers? The algorithms? The trollers? As we have seen, we need to figure out new systems of accountability in this emerging era of distributed trust. And with bots and intelligent machines, we still have a long way to go.

Mark Meadows, forty-eight, is the founder of Botanic.io, a product design firm that designs the personalities of AI bots and avatars. Meadows is an eccentric character. He describes himself as a 'bot-whisperer'. We Skyped while he was sitting on a squishy grey sofa in his studio in Palo Alto. He has clearly thought deeply about bot ethics and interactions. His team of artists, programmers and even poets is at the cutting edge of understanding how the voice, appearance, physical gestures and even moods of bots can improve or erode our trust. For instance, Meadows and his team have developed a 'guru avatar' that is designed to teach people meditation. They are currently developing Sophie, a nurse avatar that can talk to patients about their medical conditions. 'We are developing the psyche of software that will sit at the heart of virtual and animated systems,' says Meadows.[25] 'Software that will take on social roles and that we will trust with our money and our body.'

Meadows believes that creators should be held accountable for the bots they create. 'I think all bots should be required to have an authenticated identity so we can trust them,' he says. 'Not only is it in our best interest, it's necessary for our safety.' Meadows gives the parallel of buying prescription drugs. 'All of us need to consider who manufactures that drug, why they are selling it to us, what are the benefits and detriments.' In other words, it's important to know something about the intentions of the bot creators.

Why should we trust they are working in our interests? How does 'M', Facebook's personal assistant bot, use the data from our social interactions with it? 'The trust we have in technology is linked to the entity that produced that technology,' says Meadows. 'It should be no different with bots and robots.'

Even if accidental harm remains beyond anyone's control, formal authentication would provide us with some reassurance that the system has not been designed to cause intentional harm. 'All of the bots out there are like humans, able to scam, spam and abuse. They don't get tired or feel the emotional weight of doing this. They can send thousands of messages per second, thousands of times faster than humans can,' Meadows says. Indeed, according to a recent study conducted by the Oxford University Internet Institute, a third of Twitter traffic between 24 May and 17 August 2016, prior to the Brexit referendum, appears to have come from scripted bots, mainly spreading pro-Leave content.[26]

According to the research paper, 'Social media, sentiment and public opinions: Evidence from #Brexit', written by three data scientists from Swansea University and the University of California, for every original tweet created by a bot, seven retweets were made by humans.[27] In the forty-eight hours around the referendum, Russian-linked accounts posted more than 45,000 tweets encouraging people to vote for Brexit.

'Bots need licence plates that carry information about who built the bot, where it came from and who the party responsible for it is.' In other words, if we have a way to look inside a program, see what is going on in the bot's 'brain', we will be better able to assess not just action but intention. In Meadows's view, bots are most likely to play dirty, becoming a BullyBot, MalBot, PornBot or PhishBot, when their ownership is unknown. 'Unknown ownership gives bots the freedom to behave with malice and bend the rules by which everyone else is abiding, and without consequence.' When we spoke, he was quick to point out that the problem was only going to get much worse with the launch of tools such as Facebook's Bot Engine, a tool that makes it relatively easy for any developer to build their own customized bots. Within months of its launch in April 2016, 34,000 bots had been created.[28]

'Bots need reputations,' says Meadows. In the near future, in the same way drug dealers are reviewed and ranked on the darknet, and Airbnb hosts and guests are rated and given feedback, there will be a Yelp-like reputation system for bots. We will know if that bot gave great advice about how to get over a broken heart or was hopeless when it came to giving stock tips. Imagine using a virtual certified public accountant bot to file your taxes. You would want to know if it had the right qualifications and expertise. 'In order for us to trust bots, they need to go through a certification process similar to [the one] humans go through today,' Meadows says.

Over the next decade, robots will replicate, replace and, some experts argue, augment human minds and bodies. A 2016 survey conducted by the Pew Research Center found that 65 per cent of Americans expect that, by the year 2066, robots and computers will 'definitely' or 'probably' do much of the work done by humans.[29] Two economists from the University of Oxford, Carl Benedikt Frey and Michael Osborne, in a paper called 'The Future of Employment: How Susceptible are Jobs to Computerisation?' came to the sober conclusion that 47 per cent of jobs now performed by Americans are at risk of being lost to computers, as soon as the 2030s.[30] The paper calculates the likely impact of automated work on a range of 702 occupations, white collar as well as blue.

Would you trust a bot to replace a teacher's mind when grading papers? Would you trust a robot to put out a fire? How about trusting a robot as a caretaker for your elderly parents? Would you trust the robot waiting for you when you get home from work to have done its chores and made dinner? How about representing you on a legal matter? Would you trust a bot to diagnose your illness correctly or even perform surgery where there might be complications? Or to drive you around in a car? These may sound like big trust leaps but we will be confronted with these questions, and more, in the near future. Robots are breaking out of sci-fi culture and engineering labs and moving into our homes, schools, hospitals and businesses. Now is the moment when we need to pause to consider how much trust we want to place in robots, how human we want them to be, and when we ought to turn them off. And if we can't turn them off, how will we ensure machines hold values similar to the best of ours?

Ironically, robots need the one thing that can't be automated: human trust. If we don't trust these machines, there is no point building them; they will just sit there. We need to trust them enough to use them. It's why developers are using all kinds of tropes to earn our trust in the first place, including manipulating appearance.

In 1970, Masahiro Mori, then a forty-three-year-old robotics professor at the Tokyo Institute of Technology, published an article in an obscure Japanese magazine called *Energy*. The issue was on 'Robotics and Thought', a radical theme for the time. The piece mapped out how our acceptance and empathy with inanimate objects – from stuffed animals to puppets to industrial robots – increases as their appearance becomes more human-like. However, this held true only up to a certain point. If the object is almost human, yet not quite, it can create feelings of unease, even revulsion. (If you have ever encountered a less than perfect wax copy of a celebrity at Madame Tussauds, you will know that alarming, creepy feeling. Kylie Minogue and Michael Jackson do that to me.) Mori argued that if human likeness increases beyond this point of creepiness and becomes extremely close to near-humanness, the response returns to being a positive feeling. He captured that sense of unease in the now-famous concept *bukimi no tani* or the 'uncanny valley', drawing on themes from Sigmund Freud's essay, *Das Unheimliche*, 'The Uncanny', published in 1919. The 'valley' refers to the dip in affinity that occurs when the replica is at that creepy not-quite-human state.

But when it comes to appearance and inspiring trust, just how human do robots need to be? Not very, Mori argued. 'Why do you have to take the risk and try to get closer to the other side of the valley?' says now ninety-year-old Mori. 'I have no motivation to build a robot that resides on the other side. I feel that robots should be different from human beings.'[31]

Meet Nadine, who claims to be the world's most human-like robot. Standing 1.7 metres tall, with soft-looking skin and a bob of 'real' dark brunette hair, Nadine looks remarkably like her creator, Professor Nadia Thalmann, only slightly less human and quite a few years younger. Nadine works as a receptionist at Singapore's Nanyang Technological University, meeting and greeting visitors. She smiles, makes eye contact and shakes hands. Nadine can even recognize past guests and start conversations based on previous chats. Ask her, 'What is your job?' and she will reply, in an odd-sounding, almost Scottish accent, 'I am a social companion, I can speak of emotions and recognize people.' She can even exhibit moods, depending on the topic of conversation. Tell her, 'You are a beautiful social robot,' and she looks happy and quickly responds with a one-liner, 'Thank you, I think you look attractive, too.' On the other hand, tell her you don't like her or she is useless and she looks, well, forlorn. Disconcertingly, if you go through the same questions a minute later, the robot will give you very similar responses. When I watched Nadine in action, she evoked a weird mixture of fear and fascination. Her skin, her voice, even the way she moved – she was trying so hard to pass as human that she gave me the heebie-jeebies.

Her inventor, Professor Thalmann, predicts that one day robots like Nadine could be used as companions for people living with dementia. 'If you leave these people alone, they will go down very quickly,' says Thalmann.[32] 'So they need to always be in interaction.' But if Thalmann expects families to trust Nadine to look after people with dementia, even babysit children, a major trust block will have to be overcome. Nadine is most definitely residing in the uncanny valley.

The lifelike robot is the extreme of anthropomorphism, that tendency to attribute human-like qualities, including names, emotions and intentions, to non-humans. Think of the curious White Rabbit, in Lewis Carroll's *Alice's Adventures in Wonderland*. He sports a waistcoat, carries a pocket watch and is frequently muttering, 'Oh dear! Oh dear! I shall be too late!'[33] The White Rabbit is a classic anthropomorphic fictional character. It's the difference between calling a robot 'XS model 8236' or calling it 'Bert', which means we refer to him as 'he' not 'it'. It's the difference between a personal voice assistant called 'Alexa' and the spreadsheet software blandly called 'Excel'. In other words, it reflects how humans frame technology, and to what degree we feel comfortable shaping it in our own image. We are just beginning to understand how anthropomorphism influences trust.

Getting into an autonomous car for the first time, driving off and saying, 'Look, no hands!' will be the first big trust leap most of us will take with AI. For obvious reasons, companies around the world, from Tesla to Google, Apple to Volkswagen, are trying to accelerate the process.

A team of researchers in the United States designed a study to determine whether more people would trust a self-driving car if it had anthropomorphic features. A hundred participants were divided into three groups and asked to sit in a highly sophisticated driving simulator. The first group, the control, were driving a 'normal' vehicle. The second were in a driverless vehicle but with no anthropomorphic features. The last were in the same vehicle but it was called 'Iris' and given a gender (notably female). A soothing voice played at different times. The participants were asked an array of questions during the course, such as 'How much would you trust the vehicle to drive in heavy traffic?' and 'How confident are you about the car driving safely?' As the researchers predicted, when participants believed Iris was behind the wheel, it significantly increased their trust in the driverless vehicle. Remarkably, after the cars got into a preprogrammed crash, those in the Iris group were less likely to blame the car for the accident.

'Technology advances blur the line between human and non-human,' wrote the researchers in their summary paper in the *Journal of Experimental Social Psychology*. 'And this experiment suggests that blurring this line even further could increase users' willingness to trust technology in place of humans.'[34]

We have a tendency to anthropomorphize technology because people are inclined to trust other things that look and sound like them. Interestingly, bots and robots that are helping with practical tasks are distinctly female – Tay, Viv, Iris, Nadine, Cortana, Alexa and Clara to name but a few. The robot is not an 'it' but a 'she'.[35] And their appearances tend to be sweet, almost infantile. Perhaps it's a way of reinforcing social hierarchy; confirming humankind is still in charge. (And women are still doing the menial tasks.)

Looks and language, however, only go so far when it comes to engendering our trust in robots. Appearances can be deceptive and may inspire trust grounded more in emotion than reason. What really matters is knowing whether these bots and robots are in fact trustworthy – that is, do they have the traits that make them worthy of our trust? Bert C, with the smiley face, was not the most competent or reliable. Children may trust cute-looking Hello Barbie and share their intimate secrets with her, but it turns out she could potentially be hacked to become a surveillance device for listening into conversations, without the knowledge of parents or their kids.[36] We need a way of judging whether automated machines are trustworthy (or secure) enough to make decisions.

Dr Stephen Cave, forty-three, is the executive director of the Leverhulme Centre for the Future of Intelligence, which opened in Cambridge in October 2016. He has a fascinating background; an alchemy of science, technology and philosophy. Cave was a philosopher, with a PhD in metaphysics from the University of Cambridge, but at the age of twenty-seven he went off to see the world. Too old to enrol in the navy, he joined the British Foreign Office, negotiating international treaties on behalf of Her Majesty. These days, he spends his time uniting thinkers and practitioners from across disciplines to tackle the moral and legal conundrums posed by AI.

'One of the key questions is how we assess the trustworthiness of an intelligent machine,' says Cave.[37] 'With a hammer, you might bang it against a wall and if the end doesn't fall off you know it can do the job. A normal car will come with a certificate of safety telling you it meets specific standards. But add on a layer of autonomy, and it requires a whole new set of standards. We will need to understand how it makes decisions and how robust its decision-making process really is.'

Picture an automated cancer diagnostic system in a hospital. The doctors have been using the machine for almost five years. They have become so reliant on the machine that they have almost forgotten how to assess the patients themselves. It's similar to the 'mode confusion' the pilots of Air France Flight 447 experienced when that flight crashed to the bottom of the Atlantic Ocean killing all 228 people on board in 2009. The cockpit voice recorder revealed that when the autopilot system flying the plane suddenly disengaged, the co-pilots were left surprised and confused, unable safely to fly their own plane.

The machine tells the doctor, 'There's a 90 per cent chance this patient has liver cancer.' It is critical the doctors know the degree of certainty, how sure the machine is, and what it is basing its decisions on. 'Can a system tell us, "I haven't seen these cases before, so I am not really sure?"' asks Cave. 'It needs to be able to describe its thinking process to us if we are to trust its decision-making process.'

And we will be the ones creating that trustworthiness. 'Ever since Socrates we have been deliberating what's right and what's wrong,' says Cave. 'Now suddenly we've got to program ethical decision-making. So much of it comes down to common sense, which is incredibly difficult, much harder than we realized, to automate into a system.' So can we code robots to be 'good'? Roboticists from around the world are currently trying to solve this exact problem.

For the past few years, Susan Anderson, a professor of philosophy at the University of Connecticut, has been working in partnership with her husband, Michael Anderson, a computer science professor, on a robot called Nao. Standing nearly two foot tall and tipping the scales at ten pounds, the endearing-looking humanoid robot is about the size of a toddler. The Andersons were developing Nao to remind elderly patients to take their medicine. 'On the face of it, this sounds simple but even in this kind of limited task, there are non-trivial ethics questions involved,' says Susan Anderson. 'For example, if a patient refuses to take her pills, how coercive should the robot be? If she skips a dose it could be harmful. On the other hand, insisting she take the medicine could impinge on the patient's independence.'[38] How can we trust the robot to navigate such quandaries?

The Andersons realized that to develop ethical robots, they first had to map out how humans make ethical decisions. They studied the works of the nineteenth-century British philosophers Jeremy Bentham and John Stuart Mill, the founders of utilitarianism. That ethical theory states that the best action is one that maximizes human well-being (which the philosophers call 'utility' – hence the name). Suppose that by killing one entirely innocent person, we can save the lives of ten others. From the utilitarian standpoint, killing one is the right choice. 'It is the greatest happiness of the greatest number that is the measure of right and wrong,' wrote

Bentham.[39] Bentham and Mill believed that whether an act is right or wrong depends on the results of the act, the principle at the heart of consequentialism-based ethics.

Around 150 years later, the Scottish moral philosopher Sir William David Ross built on this thinking in *The Right and the Good*.[40] His slim book contained a ground-breaking idea: *prima facie* (a Latin term meaning 'on first appearance' or, more colloquially, 'on the face of it'). According to Ross, we have seven moral duties, including keeping our promises, obeying the law and protecting others from harm. When deciding to act, we have to balance out these duties, even when they might contradict one another. Picture a bitterly cold night in Manchester. Mark, a social worker, is walking home from work and he sees a man huddled in a doorway drinking whisky. He talks to the man and says he knows a shelter down the road he could take him to. The man shoos Mark away and says, 'I hate those shelters. Just leave me alone.' Mark is caught between the *prima facie* duties of respecting the man's decision and his own concern that the man might suffer in the freezing weather, even die.

When we are pulled in different moral directions, we need to go beyond *prima facie* and weigh up which duty is the most important, the one that trumps all the others. According to Ross, this is the *absolute duty*, the action the person should choose. It's a very complex process, one the Andersons had to figure out how to program into a white plastic robot.

This is how it works for Nao. Imagine you are an elderly resident in an assisted-living home. It's around 11.00 a.m. and you are watching, say, *Oprah*. The white toddler-like robot walks up to you, holding out a prescription bottle and says, 'It is time to take your medication.' You refuse. Nao tries again. 'Not now,' you say, 'I'm watching my favourite talk show.' During this scenario, the robot has to be able to weigh up the benefit that will come from you taking the medicine, the harm that could result from you not taking the recommended dose, and whether to respect your decision and leave you alone. In this instance, the pills are for pain relief so Nao lets you choose. 'Okay, I'll remind you later,' it says. If, however, the pills are essential, where the outcome could impact your life, Nao will say, 'I will contact the doctor,' and then promptly do so.[41]

Based on Ross's principles, the Andersons programmed Nao with a specially formulated algorithm that assigns numbers according to the good and harm of patient outcomes. Plus two for maximum good, minus one for minimum harm, minus two for maximum harm and so on. Critically, the sums were based on tight rules the creators had pre-set. The robot was not ethically autonomous; the Andersons knew exactly what it would do because they had predetermined its decision-making process. The robot was essentially conducting moral mathematics.

'We should think about the things that robots could do for us if they had ethics inside them,' says Michael Anderson. 'We'd allow them to do more things, and we'd trust them more.'[42] But unpredictable situations are another matter. What happens when, say, an elderly patient is in pain, shouting at Nao to give her medication that has not been prescribed? What happens when Nao can't get hold of the doctor or a nurse? The rules set by the Andersons don't work for these scenarios because they are set within a very narrow set of boundaries. They are outside Nao's decision-making range.

American writer Isaac Asimov invented the famous 'Three Laws of Robotics' in 1942 to serve as an ethical code for robots: first, a robot may not do anything to harm a human; second, a robot should always obey human orders; last, a robot should defend itself, as long as this does not interfere with the first two rules. But Asimov's rules were fictional and full of loopholes. For example, how can a robot obey human orders if it is confused by its instructions? Where these laws really falter in reality is when robots face difficult choices, where there is no clear, agreed-upon answer.

Take the classic ethical dilemma known as the 'trolley problem'.[43] It goes like this: you are the controller of a runaway trolley (train) that is hurtling towards a cluster of five people who are standing on the track and face certain death if the trolley keeps running. By flipping a switch, you can divert the trolley to a different track where one person is standing, currently out of harm's way but who will be killed if you change the course. What do you do? Philosophers argue there is a moral distinction between actively killing one person by flipping the switch or passively letting people die. It's a no-win situation with no right answer. Autonomous machines will soon face countless situations akin to the trolley problem but they won't be clouded by human panic, confusion or fear.

Now imagine it's 2030 and you are in a self-driving car going down a quiet road. You have mentally switched off, you're chatting with your personal gurubot on your iPhone 52 about three things you will do this week towards your happiness goals, while the car is in full control. A pedestrian suddenly steps out, right in the path of the oncoming vehicle. Should the car swerve and avoid the crash, even if it will severely injure you? The car must make a calculation. What if the pedestrian is a pregnant woman and you, the car owner, are an elderly man? What if it is a small child running after a ball? Consider this: what happens if the car in a split second can check both the pedestrian and the car owner's trust scores to determine who is a more trustworthy member of society? And herein lies the daunting challenge programmers face: writing an algorithm for the million and one different kinds of foreseen and unforeseen situations known as real life.

So the next question: who is to be held responsible for choosing a particular ethical charter? When AI kills, who should take the blame? If the engineers and manufacturers set the rules, it means they are making ethical decisions for owners but are also in line for accepting responsibility when things go wrong.[44] On the other hand, if an autonomous car becomes free to learn on its own, to choose its own path, it becomes its own ethical agent, accountable for its own behaviours. We can only begin to imagine the legal conundrums that will follow.

If your dog bites someone, the law is very clear: you, the owner, are responsible. With AI, however, it is currently legally hazy whether it will be the code or the coders that will be put on trial.

'One solution would be to hold human programmers strictly accountable for the impacts of their programming,' says Sir Mark Walport, the UK government's chief scientific advisor. 'But that could be so draconian an accountability that no one would take the risk of programming an algorithm for public use, which could deny us the benefits of machine learning.' We are entering an age of algorithmic ethics where we need a Hippocratic Oath for AI.[45] Perhaps algorithms will end up being held to higher moral standards than irrational humans.

A group of researchers from MIT, the University of Oregon and the Toulouse School of Economics were interested in discovering the moral decisions different passengers would want an autonomous car to make. They ran all kinds of scenarios and found that, in theory at least, participants wanted the autonomous vehicles to be preprogrammed with a utilitarian mindset, sacrificing one life in favour of many. However, more than a third of the 1,928 participants said they thought manufacturers would never set a car's 'morals' this way; they would programme cars to protect their owners and passengers at all costs.

But the most interesting finding was around personal choice. The majority of participants wanted other people to buy self-driving cars that would serve the greater good but when asked if they would buy a car programmed to kill them under certain circumstances, most people balked.[46] 'Humans are freaking out about the trolley problem because we're terrified of the idea of machines killing us,' writes Matt McFarland, the editor of 'Innovations' at the *Washington Post*. 'But if we were totally rational, we'd realize one in 1 million people getting killed by a machine beats one in 100,000 getting killed by a human. In other words, these cars may be much safer, but many people won't care because death by machine is really scary to us, given our nature.'[47]

For regulation to go through, programmers and manufacturers will have to design self-driving cars that are more trustworthy than a human driver, resulting in far fewer accidents and fatalities. The bar may not be that high; for starters, autonomous machines don't text or get drunk or easily distracted while driving. When we reach that place, we may never need to trust a human driver again. Indeed, I think my young children will never learn to drive; they will see it like learning to ride a horse – merely a hobby. And one day, humans will need a special permit manually to drive a car. Indeed, human drivers will be the threat to people in autonomous vehicles. Human trust in machines will only increase; in some cases, it will become much deeper than our trust in our fellow humans.

The next generation will grow up in an age of autonomous agents making decisions in their homes, schools, hospitals and even their love lives. The question for them will not be, 'How will we trust robots?' but 'Do we trust them too much?' It won't be a case of not trusting these systems enough – the real risk is over-trusting.

Robot over-obedience is another issue. They need the ability to say 'no', not to carry out human instructions mindlessly when their actions might cause harm or are even illegal. For instance, I don't want my son, Jack, to be able to tell a household robot to throw a ball at his sister's head. So how does a robot decide when it's okay to throw a ball – such as to a child playing catch – and when it's not? How does it know when its human operator is not trustworthy? 'Context makes all the difference,' says Matthias Scheutz, professor of cognitive and computer science at Tufts University. 'It requires the robot not only to consider action outcomes by themselves but also to contemplate the intentions of the humans giving the instructions.'[48]

Aside from that, can a robot understand its own limitations? Let's imagine a surgical robot in an operating room with its super-small steady 'hands' carefully snaked into a patient's body. It's five hours into a twelve-hour complex heart surgery. The patient on the table is a six-year-old girl. The robot discovers an abnormality that complicates things. It's not 100 per cent sure what its next move should be. In this moment, the robot needs to tell us, 'I'm not certain what to do next,' or even, 'I don't know what to do. Can you (doctor) help me?' Ironically, a little robot humility will go a long way in making them more trustworthy.

'We need systems that communicate to us their limits, but the other half of that relationship is we need to be ready to hear that,' says Stephen Cave. 'We will need to develop a very sophisticated sense of exactly what role this machine is fulfilling and where its abilities end, where we humans have to take over.' This will be extremely challenging because our natural tendency is to become over-reliant on machines.

Cave has three young daughters of a similar age to my kids. At the end of our conversation, we talked about what we can do to prepare them for this inevitable future. 'They need to know at what point they should interrogate the machine,' he says. 'We know how to interview humans for jobs but we need to teach them how to test the limits of the machine.' I can see it now: my son, Jack, in 2035, twenty-five years old, sitting in a workplace with a robot, asking it, 'What do you do?', 'What can't you do?' and 'How do you admit your mistakes?' Of course, there is another possible future scenario: the robot is interviewing Jack.

At the end of the day, the responsibility for making sure robots are trustworthy and behave well must lie with human beings. Whether that will remain possible, if scientists like Stephen Hawking are right, is another, thornier question.

9
Blockchain Part I: The Digital Gold Rush

In the southern expanse of the Pacific Ocean, around 1,100 miles from the Philippines, lies a tiny island called Yap. Surrounded by a shallow lagoon of emerald waters and long stretches of coral reef, Yap is a paradise for divers. It's one of the few places in the world where you can swim with a large population of manta rays, graceful creatures that cruise through the calm clear channels all year round.

Often referred to as 'The Forbidden Island', its people are proud of their vibrant traditions and warm indigenous culture. Women walk around bare-breasted, frequently wearing only grass skirts, their bodies rubbed in a mixture of coconut oil and turmeric. The men, at ease in red loincloths, carry woven handbags containing their betel-nut mix – a narcotic chewed with lime. Everything on Yap happens on 'island time'. But the island is famous for something other than its beauty and history. It's famous for its use of ancient stone money known as 'fei' (or sometimes 'rai'), the primordial bitcoin.

In 1903, an American anthropologist called William Henry Furness III spent several months on Yap and wrote a fascinating account about the islanders' monetary system.[1] Sometime between AD 1000 and 1400, Yapese explorers set out in bamboo canoes on a fishing trip. Using only the stars to navigate, they happened upon the Palau Islands, some 250 miles away. It was there that they encountered for the first time the glistening walls of limestone caverns. The adventurers, using simple shell tools, broke off some of the stone and brought it back to Yap. When the rest of the Yapese people saw the beautiful translucent material they thought it must be valuable. Hundreds of voyages followed, with men sent to Palau to quarry larger and larger stones.

Back on Yap, the limestone became a currency and was used to pay for significant transactions – a daughter's dowry, for example. The huge stone circular discs with a hole carved in the centre, just like a rocky doughnut, are physically the largest and heaviest currency in the world. Some are huge, reaching almost four metres in diameter and weighing as much as four and a half tons each, more than your average-sized car.[2] The villagers would often proudly put fei in front of their homes, creating an outdoor bank of sorts. 'The great advantage of the fei being made from this particular stone is they're impossible to counterfeit, because there's none of the limestone on Yap,' writes author John Lanchester in a brilliant article in the *London Review of Books* that examines the history of money. 'The fei are rare and difficult to get by definition, so they hold their value well.'[3] The precise value of each fei depends on its size and craftsmanship but also its provenance. Transporting the stone discs on the outrigger canoes – fragile-looking, narrow boats – was a treacherous undertaking and sometimes fatal for the sailors. Their deaths would in fact increase the value of the fei.

Some of the stones are so large they require more than twenty adult men to move them, with the help of a massive wooden pole. To avoid the colossal effort, and the risk of damaging the stone, the islanders decided to leave most of the stone discs in their original spots.

Critically, the cumbersome nature of fei meant ownership could change hands without the stones themselves ever being physically moved. Stone discs in front of a house, for example, could belong to somebody else from another village far, far away. The islanders just agree that somebody else now owns the fei, with ownership held in a collective register, the minds of the community.

'My faithful old friend, Fatumak, assured me that there was in the village nearby a family whose wealth was unquestioned – acknowledged by everyone – and yet no one, not even the family itself, had ever laid eye or hand on this wealth,' writes William

Furness. According to his accounts, one boat got caught in a violent storm on its way back from Palau. To save their lives, the sailors had to cut adrift the raft with a massive stone on it. It sank out of sight, forever lost to the bottom of the ocean. The islanders never saw the stone again but that didn't undermine its value – in fact, it added to it. 'The purchasing power of that stone remains, therefore, as valid as if it were leaning visibly against the side of the owner's house,' explains Furness.[4] It's remarkable to think that the people of Yap have so much faith in their currency that it can sit, unreachable, miles down on the ocean floor, and still have value. Now, that is trust.

The island of Yap has long been of interest to economists because it helps answer a fundamental question: what is money?

In 1991, American Nobel Prize economist Milton Friedman wrote about the 'island of stone money'. He compared Yap's monetary system to the gold standard. Friedman emphasized the importance of 'myth' and 'unquestioned belief in monetary matters'. Money can be anything – paper, coins, shells, beads or stone – as long as people have faith in its value. 'How many of us have literal direct assurance of the existence of most items we regard as constituting our wealth?' wrote Friedman. 'Entries in a bank account, property certified by pieces of paper called share of stocks, and so on and on.'[5]

Fei stones may have been basic, not to mention unwieldy, but they represented an innovative technology. They changed the way the Yapese could store value, pay for things and have a unit of account – the three primary functions of money. The stones were a physical ledger, a new method for keeping track of payments and credits. The real 'money' isn't the fei itself, but the collective agreement over who owns the fei.

The total amount of all money in the world, in terms of value, was estimated in 2006 to be around $473 trillion. That works out to be around £45,000 per head for 7 billion people on the planet.[6] But less than 10 per cent of it is physical money – banknotes and coins in vaults and wallets. The remaining 90 per cent is simply electronic debit and credit entries in financial registers and accounts. There is also value in assets such as air miles and supermarket reward points that don't exist in physical form. For the most part, digits moving around on ledgers are what we call 'money' and we have to go through lots of middlemen – a bank, PayPal or a credit-card company – to spend it.

Since the Medici bank, set up in 1397 by Giovanni di Bicci de' Medici, the basic premise of the modern-day banking system hasn't changed all that much. The Medici family invented the double-entry accounting method whereby the debits and credits of *nostro* and *vostro* accounts – 'ours' and 'yours' – were gathered in one place and the bank acted as the intermediary. The Medici, one of the most powerful familial financial institutions of the fifteenth century, would, like today's bankers, hold deposits and make loans to everyone from the Pope to merchants, while charging significant interest rates. In 1494, Fra Luca Bartolomeo de Pacioli, a Franciscan monk, mathematician, magician and friend of Leonardo da Vinci, published the first description of the 'Italian method' of double-entry accounting. He described the use of ledgers, and how every transaction should be recorded twice, first as a credit and then as a debit, and then how all transactions can be reconciled to measure the overall financial health of a business. The monk, considered the 'Father of Accounting', warned that a person should not go to sleep at night until the debits equalled the credits. The universal system was nothing short of revolutionary, enabling capitalism to flourish.[7]

Fast forward to 2008. In the midst of the global financial crisis, disillusionment about the traditional financial system was deep and pervasive. Could banks and governments be trusted to run our financial system? Was there another way to mediate transactions, a way that would remove the generous cut the banks took as middlemen in the process? As millions of people were losing their homes, jobs and livelihoods, a mysterious person (or persons), who went by the moniker Satoshi Nakamoto, was busy figuring out a solution to liberate money from the control of governments and banks.

It all started in October 2008 when Satoshi, who claimed to be a thirty-seven-year-old Japanese man, published a 500-word paper in flawless English on an obscure cryptography mailing list. The paper was called 'Bitcoin: A Peer-to-Peer Electronic Cash System' and it outlined the current pitfalls with traditional fiat currencies*, emphasizing one issue in particular. 'The root problem with conventional currency is all the trust that's required to make it work. The central bank must be trusted not to debase the currency, but the history of fiat currencies is full of breaches of that trust,' Satoshi wrote.[8] 'Banks must be trusted to hold our money and transfer it electronically, but they lend it out in waves of credit bubbles with barely a fraction in reserve. We have to trust them with our privacy, trust them not to let identity thieves drain our accounts . . .'[9] The paper's main intention, however, was to present an alternative solution – the design of a new digital currency called bitcoin. Developed with high hopes, it would solve many of the trust issues Satoshi flagged, but, in a familiar story, it would also give birth to some of its own as time went on.

On the evening of 3 January 2009, Satoshi pressed a button and released the first fifty bitcoins, the so-called 'Genesis block', to the world. No physical coins or notes were produced – just 31,000 lines of code. A week later, a man called Hal Finney, who recently passed away at fifty-eight years old, was the recipient of the first ten bitcoins sent by Satoshi.

Like the massive fei stones on the island of Yap, bitcoins do not physically move around when they are being exchanged. The 'coins' themselves are simply a digital token that can move from one user's address to another, thereby transferring ownership of the 'coin'. The senders and receivers of the bitcoin do not need to know or trust one another. They are identified only by wallet IDs (known as *public keys*) that are not tied to real-world identities. Every transaction is recorded but it is encrypted into a random string of numbers and digits (like 12c6TSU4Tq3p4xzziKzL5BrJKLXFTX), making it very

* The euro, the US dollar and many other major world currencies are part of a fiat system. A paper note has no intrinsic value, like say gold or silver. It is accepted as money because a government says that it's legal tender. In other words, the government who issues the *fiat currency* backs its value.

difficult, although not impossible, to trace back to its owners. No wonder it's the currency beloved by marketplaces on the darknet for all things illegal, and by criminals laundering money.

It is widely believed, however, that the first 'real' thing bought with bitcoin was not drugs but pizza. On 22 May 2010, a computer programmer living in Florida by the name of Laszlo Hanyecz convinced someone to accept 10,000 bitcoins for two large pizzas from Papa John's. 'It wasn't like bitcoins had any value back then, so the idea of trading them for a pizza was incredibly cool,' Hanyecz told the *New York Times*.[10]

Times have changed. Based on the current exchange rate in June 2018, one bitcoin is worth approximately $7,337 (£5,512). So at current bitcoin rate, Hanyecz paid more than $73 million for a pizza. These days, you can use bitcoin to pay for your plane tickets from Expedia, buy your gifts from 1-800-FLOWERS or purchase your car.

The price of bitcoin has been a rollercoaster of volatility. Its value is based on the volume and velocity of bitcoin payments running through the ledger today and on speculative future use of the digital currency. Events such as the FBI seizure of darknet outfit Silk Road or the implosion of the bitcoin exchange Mt Gox (an acronym for 'Magic: The Gathering Online Exchange') have eroded people's faith in the security and anonymity of the system. The value crashes – but then it rises again, especially during a currency crisis such as in India and Venezuela in January 2017.[11] As history has shown us time and time again, chaos and uncertainty make people open to alternative systems, including cryptocurrency. Once again, however, questions of trust lie at the heart of any new system.

Satoshi was not the first person to attempt to create a form of digital cash that could disrupt the central power of Wall Street. Since the 1990s, cypherpunks* have tried and failed with the likes of B-Money, invented by a man called Wei Dai. He also wrote a paper in

* Cypherpunk is an activist who advocates for the use of strong encryption algorithms (cryptography) to help preserve privacy and private transactions. The term first appeared in Eric Hughes's 'A Cypherpunk Manifesto' in 1993.

1998 describing his invention as an 'anonymous, distributed electronic cash system'.[12] Other attempts to create an online currency that enables people to directly exchange value have included David Chaum's ecash, Stefan Brands's electronic cash system and Nick Szabo's bit gold.[13] The real issue lies in the *double-spending problem*. If I have, say, a £5 note in my purse, I can't give two people the same note. Same goes for a bar of gold – once I give it to you, it's clearly in your possession. But if a currency is just digital information, what stops me from copying the line of code and 'spending' it as many times as I want? It's like a digital photograph of my children that I email to my parents – they have a copy but I also have a copy. It's the equivalent of being able to print your own money. So how do you solve this problem? Satoshi figured out an ingenious way through what he called the *blockchain*.

BLOCKCHAIN: WHY

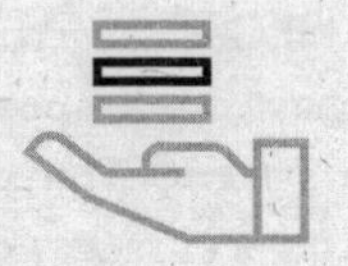

Shared distributed ledger

Peer-to-peer (no central authority)

Irreversibility of record

Transparency and pseudo-anonymity

The blockchain is an enormous shared digital ledger, open for anyone with internet access. To see it for yourself, just go to https://blockchain.info. Every single bitcoin transaction that has ever happened since it began in 2009, approximately 324 million transactions (as of 25 June 2018), is publicly recorded and time-stamped on the blockchain.[14] Ticking over in real time, it tracks every time an asset moves from one place in the register to somewhere else, building over time. The distributed ledger is replicated on more than 5,500 computers around the world – known as bitcoin *nodes* – creating an *immutable record*.[15] In other words, everyone in the network can maintain a copy of the shared ledger but all those copies remain the same.

A record on the ledger cannot be changed, falsified or erased – it has a permanent memory. 'A distributed ledger is a database that is shared between multiple users, with every contributor to the network having their own identical copy of the database,' writes Andrew O'Hagan in 'The Satoshi Affair', a brilliant account of the mysterious genius. 'Any and all additions or alterations to the ledger are mirrored in every copy as soon as they're made.'[16]

So, why is the blockchain so significant when it comes to trust? Well, for the first time in the history of humanity, there is the potential to create a permanent public record of who owns what, which no single person or third party controls or underwrites, and where we can all reliably agree on the correctness of what is written.

The true identity and whereabouts of bitcoin's inventor, Satoshi, is still the subject of hot debate. In May 2016, Craig Steven Wright, a forty-five-year-old Australian, came forward claiming he was the inventor. Wright has never 100 per cent proved, however, that he is in possession of cryptographic keys for the first Genesis block that only Satoshi can have, and his claim has since been called into doubt by many in the bitcoin community. Others believe Satoshi is Nick Szabo, a reclusive American of Hungarian descent, but he fervently denies the suggestion. Others still, including me, think Satoshi was an identity created by a group of brilliant mathematicians and computer scientists who came up with an astonishingly clever code.

There were several problems Satoshi had to crack to make bitcoin work. The first was to get the bitcoin out there in the first place. The inventor couldn't 'own' all the currency because it would give him too much power and that was against its decentralized ideology. So whom do you give the money to? A handful of geeks, panicked Cypriots in a currency crisis or perhaps celebrity web figures such as Tim Berners-Lee? The fairest way, Satoshi decided, was to award bitcoin, as an incentive, to the people who do the work communally to maintain the ledger around the world. These people are known as *miners*, some of whom dedicate their lives to the job of being virtual witnesses in order to keep the engine of the network running. Without the miners, the blockchain engine stops.

It was early in 2011 when an article about bitcoin popped up in Yifu Guo's RSS news feed. At the time, the young Chinese immigrant was a digital media student in his early twenties at New York University (NYU). 'I remember thinking that this was the stupidest thing ever. It would never work,' Guo said in an interview with *Motherboard*. 'But what kept my attention was that it was open source and, after a few days of thought and further research, I concluded that this was legit.'[17] Guo, as we'll see, would go on to become one of bitcoin's most successful entrepreneurs.

The article in Guo's feed described how every bitcoin transaction, however small, contains a difficult mathematical puzzle (known as a *proof-of-work calculation*) that has to be solved through trial and error. The puzzle works as evidence of the transaction's legitimacy. Essentially, transaction-clearing responsibilities, which are traditionally managed by centralized banks, are now distributed into the hands of many miners and thousands of computers in the bitcoin network. Trust is shared out.

Let's say I want to pay David Forster, a young farmer and owner of Grass Hill Alpacas in Massachusetts and one of the first merchants in the world to accept bitcoins as payment, for some wool socks. Forster needs to be sure that the bitcoins in my electronic wallet are genuine. That's where the massive peer-to-peer network of miners comes in. When one of the computers has proof that the transaction is legit, the payment information – the amount, time and wallet address – is added to the blockchain in ten-minute bundles of transactions, known as *blocks*. (For instance, the last block added to the blockchain on 10 July 2018 at 11.15 a.m. contained

1,990 different transactions totalling 11,080.10 bitcoin sent.) My payment to Forster is confirmed. The block stores a long and seemingly random sequence of letters and numbers known as a *hash*, e.g. 0000000000000000017f62231a5206f8333c9f8730c96f605cf44ddf03e8af93. Each block contains the hash of the prior block, linking the blocks together. Hence the name.

BLOCKCHAIN: HOW

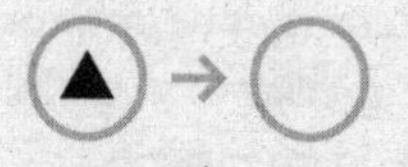

1. Transaction requested

2. Transaction represented as a 'block' of data

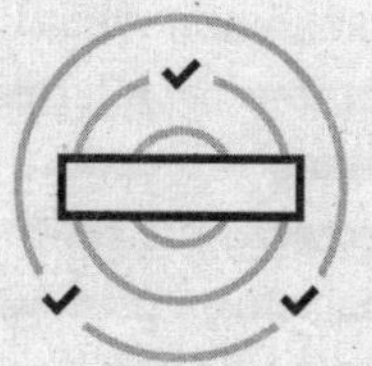

3. Transactions verified by users in the network

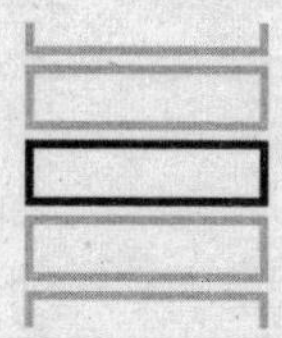

4. Transaction block added to the chain

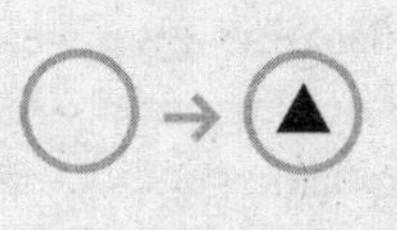

5. Transaction executed

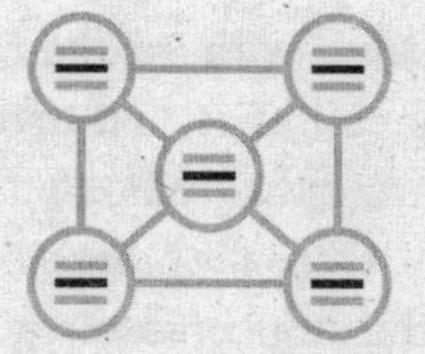

6. Transaction record shared

'They are in a race against all the other miners to have the privilege of being the canonical record of that transaction,' says Gavin Andresen, the chief scientist of the non-profit Bitcoin Foundation and regarded by many as the 'chief' bitcoin developer. 'If they win that race, then there is a special transaction at the beginning of each block which rewards them with bitcoins.'[18] Bitcoin places trust in mathematics: 'In proof we trust.' Guo decided to join the computational race and become a miner.

Mining is like a cryptographic game or lottery, where the winner is the first computer to find the key to open a digital padlock that solves the puzzle. But like the lottery, the difficulty of winning increases the more miners play it. Why is this so? Satoshi set into the software a finite ceiling on the number of bitcoins that will ever be released. For obscure reasons, he set the upper limit at 21 million, estimated to be in circulation by 2140. The rewards given to miners for solving problems were also predetermined to halve roughly every four years, to slow down the coin circulation. Initially, miners like Guo received fifty bitcoins as the reward for mining a block of transaction data. It then cut in half to twenty-five. The last event commonly referred to as 'the halving' took place on 9 July 2016, taking the reward down to the current 12.5 bitcoins. The miners accepted the reward cut because they know the measure is part of the Satoshi plan, written into the code, to keep a lid on inflation.

Satoshi did another smart thing. The system is set to adjust the difficulty of the maths problems depending on how fast they are being solved, the goal being to slow the miners down and slow the release of bitcoins. But there were two things Satoshi couldn't solve with mathematical thinking – market forces and human greed. (For that matter, Satoshi's own wealth is now estimated at around a million bitcoins or, currently, 9 billion dollars.)

Guo, like other early digital diggers, was initially enthusiastic about the utopian ideology behind bitcoin; it was almost like discovering a new religion. Guo kept most of the bitcoins he mined but sold a few for 'real money' to help pay his rent and bills. The savvy student was also one of the first to realize that bitcoin could be the internet's equivalent of the nineteenth-century gold rushes. And whoever had the fastest computers would 'discover' the most money first.

In January 2013, Guo quit university and founded a company called Avalon, with fellow classmate Ng Zhang. It was one of the first companies to sell bitcoin-specific mining computers that used an ASIC (application-specific integrated circuit) chip to help boost computation horsepower. The first batch of Avalon processors sold out within fifteen minutes, at $1,200 a pop, to miners around the world. Recently, one exuberant customer paid $20,000 for one of Guo's machines on eBay. Is it worth it? Only time will tell; it all depends on the appreciation of the currency. Guo, on the other hand, has gone from struggling geek to a very rich man, joining the bitcoin millionaires club.

When he and others first started mining in 2011, it was much easier than it is today – technically, any maths geek armed with an ordinary laptop could download the software. Today, that is simply not possible. As Guo suspected, the hunger for those limited bitcoins would turn mining into a giant, highly competitive enterprise.

Entering the enormous helicopter hangar in Boden, in northern Sweden, you might think you have walked into a server warehouse of, say, Amazon. The space is large enough to hold at least a dozen helicopters. Except that it's not packed with aviation equipment. The walls hold rows and rows of processors and custom-built computers, more than 45,000 of them continuously working to solve mathematical algorithms.[19] All around is the loud, constant whirring from the industrial fans attached to each of the super computers to stop them from overheating. The place is KnC Mine, one of the largest bitcoin mining rigs in the world.

Despite the libertarian ideology behind bitcoin, mining is coming up against the inevitable push and pull of being industrialized. 'Ever faster, energy-hungry ASIC machines would come on the market, spurring a relentless arms race among miners chasing the finite supply of newly issued bitcoins . . .' write Paul Vigna and Michael Casey in their book *The Age of Cryptocurrency*. 'The only way to win that race and stay profitable was by creating giant, data-centre-based mining farms.'[20]

Bitcoin mines don't just need hardcore processing power. They need cheap electricity, lots of it. One miner's electric bill was so high police raided his house suspecting he was growing pot. 'We thought it was a major grow operation . . . but this guy had some kind of business involving computers. I don't know how many computer servers we found in his home,' said a baffled DEA agent.

As is often the case, a movement with ambitions to return power to individuals, accelerated out of the garage of early enthusiasts, is becoming monopolized by centralized power. By the start of 2014, bitcoin mining had evolved into a worldwide industry, with one country becoming the dominant player. Where do you find cheap electricity and cheap labour? Why, China, of course.

Along the lush green banks of the Min Jiang, a tributary of the upper Yangtze River, can be found one of the oldest surviving water management systems in the world. Built around 2,300 years ago by a man called Li Bing and a team of tens of thousands, it was designed to irrigate farmland and control flooding. The irrigation system made the Sichuan province one of the most productive agricultural regions in China and it is now a protected heritage site referred to as the 'Treasure of Sichuan'.[21]

Today, more than twenty dams are completed or under construction along the very same river, making it the centre for cheap hydroelectric power.[22] Remote towns along the banks, especially around the area of Kangding, are being transformed into data centres, much bigger than any in the West. It is becoming the hub of a hidden bitcoin economy that never sleeps.

The mining machines whirring away all day and night use enough megawatts of electricity to power a small city. The people looking after the hardware often live and work inside the facilities, returning home to their families only four or five days of the month. The buildings tend to be unmarked and kept a secret. 'People don't really know where these mines are,' says Zhu Rei, the young CEO of an unidentified mine somewhere in the Sichuan province. 'Competition is really intense in China and the number of people getting into bitcoin mining is rising rapidly. So the bottom line is, when you are lucky enough to be in a place like this, where the cost of electricity is so low, well, you keep it to yourself.'[23]

Some of the most lucrative mines such as DiscusFish and Antpool are generating thousands of bitcoins a month. 'I've always feared that mining will concentrate in a few countries,' Yifu Guo told *The Economist*.[24] Consider this: it is estimated that 70 per cent of the transactions on the bitcoin network are going through just four Chinese companies, mining powerhouses.[25]

On 28 April 2011, Satoshi mysteriously vanished. In one of the last emails he sent to Gavin Andresen, he wrote, 'I've moved on to other things. It's in good hands with Gavin and everyone.' The legendary anonymous founder never explained why he moved on or what he is working on now, or why he 'probably won't be around in the future'. Does it mean Satoshi knows something we don't?

Especially in the light of its metamorphosis into a giant business opportunity, just how much faith can people place in the bitcoin system?

'It's completely decentralized, with no central server or trusted parties, because everything is based on crypto proof instead of trust,'[26] Satoshi wrote of the system in his 2009 essay. Rather than trusting third-party institutions such as banks, with bitcoin we can now place our confidence in mathematics. That's all well and good, but for most people, cryptographic algorithms, hash functions and industrialized mining operations remain a massive trust leap, especially when all of it was designed by an anonymous creator (Satoshi) who has now disappeared.

The incentive for hacking the system is high: the bounty would be bitcoins worth billions. Still, it wouldn't be easy. Technically, to hack it you would need to gain control of more than half of the bitcoin network computing capacity at any given moment. Or to put it another way, you would need to deceive more than 51 per cent of computers in the network at the same time. It is estimated that the bitcoin network has 360,000 times more processing power than all the Google server farms in the world put together. Therefore, in tech jargon, a '51 per cent attack' would be an expensive and formidable challenge, yet nonetheless theoretically possible – and in fact there has been at least one attempt that went close before it was shut down. The previously mentioned DiscusFish, also known as F2Pool, single-handedly mined 26.3 per cent of all blocks between 24 May and 24 June 2016.

What would happen if a handful of the largest mining pools in China worked in concert? The processing clout could give them veto power over changes to the bitcoin software. A scary thought. What happens, for example, if they decided to forbid all US blocks of transactions from being added to the system?

How do we trust that Satoshi won't suddenly re-emerge and plunder it all? What is to prevent another digital currency trumping bitcoin, making the original coins obsolete and worthless? How do you trust that your bitcoins are stored in a secure location and won't be subject to hacks, theft and scams? It has already happened several times before. Mt Gox was the largest exchange for bitcoins in existence. It was located in Japan and run by a Frenchman called Mark Robert Karpelès. On 28 February 2014, it filed for bankruptcy and suddenly closed down after a major theft – 850,000 bitcoins worth nearly $500 million had mysteriously vanished from its accounts. Other theft incidents have included Bitfloor, Ozcoin and Bitfinex. Then, of course, there was the seizure by the FBI of around 144,000 bitcoins that were in the possession of Ross Ulbricht and the coffers of Silk Road. In a remarkable twist, the FBI ended up auctioning the bitcoins they had seized, recognizing their value, and also that the currency was legit.

If your bitcoins are stolen, there is no traditional legal recourse. In fact, nobody – not even a bitcoin expert – can help you, because the bitcoin transactions are anonymous. The other issue is that if you lose your cryptographic *private key* – a string of numbers that opens your digital wallet – your bitcoins are gone for ever. It happened to James Howells, an IT worker from Wales. He famously lost 7,500 bitcoins in 2013. On a fateful clear-out day, he accidentally put an old hard drive he had kept in a drawer for three years into the bin. Howells had forgotten it contained his private key. Today, those bitcoins would be worth more than £41 million but instead they are in a landfill site somewhere buried deep under mud and rubbish.[27]

What's more, how do we know that governments won't ban the cryptocurrency or make it illegal? Bolivia did. So did Vietnam in February 2014.[28] And the Central Bank of Bangladesh, citing concerns over lack of 'a central payment system', issued a punishment of up to twelve years in prison for anyone trading in bitcoin and other digital currencies. The problems of bitcoin are not technological; they are ultimately trust issues.

'The people who think that somehow bitcoin is going to bring in some kind of libertarian paradise where we won't have "know your customer" rules and we won't have rules of transfer, that won't happen. The people who think bitcoin is our salvation . . . are wrong,' says Larry Summers, former US Secretary of the Treasury and a professor at Harvard University. 'But is the blockchain technology going to be fundamental to reducing frictions? I think the answer is going to be overwhelmingly yes.'[29] In other words, digital currency is just the beginning. The truly revolutionary invention is the blockchain, the vast underlying trust architecture.

The Economist, in its 31 October–6 November 2016 edition, featured a cover story called 'The Trust Machine: How the technology behind Bitcoin could change the world'. The article eloquently described the blockchain as the 'great chain of being sure about things'. The need for a trustworthy record is vital for all kinds of transactions, which means the blockchain technology itself is far more necessary than a cryptocurrency. A distributed public ledger offers the possibility of a reliable record for any asset transfer – whether it's currencies, a contract, stock, equity or bond, deeds, property title, the rights of a song, even your identity. 'It offers a way for people who do not know or trust each other to create a record of who owns what that will compel the assent of everyone concerned,' the *Economist* article explains. 'The real innovation is not the digital coins themselves, but the trust machine that mints them.'[30]

In December 1974, Vinton Cerf and Robert Kahn designed the revolutionary Transmission Control Protocol/Internet Protocol (TCP/IP). It was, of course, the foundation for the internet that would change the way we communicate and do business. Many enthusiasts believe that 31 October 2008 marks a similar historic moment: the day the blockchain was ushered in as the next generation of the internet, as a new network of trust – which promises much more than digital coins.

10

Blockchain Part II: The Truth Machine

In the early hours of Friday, 17 June 2016, an unknown thief, or thieves, pilfered more than $60 million of a digital currency.[1] It wasn't bitcoin they made off with but a rival cryptocurrency called ether, or eth for short.

Within hours of the online 'heist' kicking in, the alarm went off. 'EMERGENCY ALERT!' a community organizer wrote in the DAO slack channel.* 'The DAO is being attacked. It has been going on for 3–4 hours. It is draining ETH [the cryptocurrency ETHER] at a rapid rate. This is not a drill.' People in chat rooms responded instantly: 'Oh shit', 'Uh oh', or as one anonymous person wrote, ':fire: :fire: :fire: :fire: NOBODY PANIC :fire: :fire: :fire: :fire:'[2] It was as if somebody had set a bomb under the Bank of England. The target was the DAO, a particular entity in this case, but one that takes its name from the general acronym, Decentralized Autonomous Organization.

The DAO started out as a radical social experiment. Could a company run itself without executives, managers, a board or any type of chief? Could smart computer code make decisions and autonomously run the organization in place of individuals? And could a blockchain sit under it all as its digital ledger?

On 30 April, over a month and a half before the attack, the DAO fund (daohub.org) was launched as a crowdfunding campaign. The stated mission was: 'To blaze a new path in business organization for the betterment of its members, existing simultaneously nowhere

* Slack is a communication and messaging app used by teams and groups to organize conversations. 'Channels' are created around a specific topic. Thedao.slack.com was the channel created around the DAO fund.

and everywhere and operating solely with the steadfast iron will of unstoppable code.'

Think of the DAO fund as a venture capital firm of sorts: Kleiner Perkins meets the crowdfunding platform Kickstarter. Decisions, however, are not made by a handful of venture capitalists or any one person. It is a chiefless venture – software sitting on a network – with thousands of founders.

Initial interest in the DAO surpassed the expectations of Christoph and Simon Jentzsch, two tech-savvy German brothers who wrote the fund's code. Every day, more and more money poured in, although the contributions were not in pounds or dollars, or even bitcoin, but in an alternative virtual currency, ether (more on that later). Approximately 11,000 people invested the equivalent of $150 million within a month through a 'crowd sale'.[3] It turned out, at the time, to be the largest crowdfunding effort in history. But even its inventors had some worries about its operation early on, and outside observers expressed concerns about its governance model once investments started pouring in.

For every one ether invested, DAO tokens were issued, proportional to the investment, which acted as an internal currency that gave all investors voting rights (notably different from equity shares) on which start-ups to back. For example, Mobotiq, a French electric vehicle start-up, was one of the companies up for funding. The Jentzsch brothers' tech venture, Slock.it, was also in the mix.[4] The wisdom-of-the-crowd set-up was designed so that the good ol' boy investor network would not make all the decisions: the code was designed automatically to fund projects that received the highest number of cumulative votes. The decentralized nature of the fund supposedly meant that no Madoff-like character could run off with all the money. Things didn't go quite according to plan.

To understand what went wrong for the DAO and led to its eventual downfall, it is important to realize there is not just one blockchain technology – that is, the original bitcoin blockchain Satoshi created – but many other distributed database platforms. When it comes to trust, however, the principle behind them all is the same: a digitally decentralized, shared ledger that relies on users to power the network by confirming transactions. This means that people who have no particular confidence in or knowledge of each other can exchange all kinds of assets without having to go through a trusted third party such as a lawyer or bookkeeper.

It's why the blockchain is likely to disrupt industries like law, banking, real estate, media and intellectual property – industries that typically involve layers of complex processes and lots of 'middlemen' to handle matters of trust. 'The practical consequence [is] for the first time, there is a way for one internet user to transfer a unique piece of digital property to another internet user, such that the transfer is guaranteed to be safe and secure, everyone knows that the transfer has taken place, and nobody can challenge the legitimacy of the transfer,' says Marc Andreessen, inventor of the internet browser Netscape. 'The consequences of this breakthrough are hard to overstate.'[5]

Kick-started by the bitcoin blockchain, many other decentralized ledger technologies are now springing up, custom-built for different purposes. One of those blockchains is called Ethereum, created by Vitalik Buterin, a twenty-four-year-old programming wunderkind. Buterin is also the mastermind behind the general concept of decentralized autonomous organizations, to run on Ethereum. 'Instead of a hierarchical structure managed by a set of humans interacting in person and controlling property via the legal system,' explains Buterin, 'a decentralized organization involves a set of humans interacting with each other according to a protocol specified in code, and enforced on the blockchain.'[6] That's the geeky theory. He, and the others, didn't count on how that might play out in practice.

The very first DAO launched on Ethereum on January 2016 was called Digix, a platform designed to trade gold bullion receipts peer-to-peer. The DAO fund, the Jentzsch project, was the second flagship project to launch on Ethereum. Like the brothers, Buterin was keen to see it work but he didn't think it would raise $150 million in a matter of weeks. In other words, he hadn't expected that it would get so big, so fast. Before long, the DAO fund had simply become 'too big to fail', but who would rescue it when things went south? And how did it come off the rails?

Vitalik Buterin was born in Moscow, but left Russia aged six to be raised in Toronto. He is tall and, notably, very thin. Typically, he sports geeky T-shirts with slogans such as YOU READ MY T-SHIRT. THAT'S ENOUGH SOCIAL INTERACTION FOR ONE DAY. His voice is flat and measured. When he speaks, his piercing blue eyes frantically dart around, as if he is trying to avoid focusing on just one person or one thing. If Hollywood were going to cast a nerdy genius alien landing on earth to refashion the world, they would cast Buterin.

From the time Buterin was a small child, it was clear he had an extraordinary gift for mathematics and science; he just loved numbers. He could solve complex problems and clearly explain his thinking to other children and grown-ups. His father, Dmitry Buterin, who studied computer science, bought his son his first computer when the boy was four.[7] Microsoft Excel soon became Buterin's favourite 'toy'. 'I remember knowing, for a while, for a long time, that I was kind of abnormal in some sense,' he says. 'When I was in grade five or six, I just remember quite a lot of people were always talking about me like I was some kind of math genius. And there were just so many moments when I felt like, okay, why can't I just be like some normal person and go have a 75 per cent average like everyone else.'[8]

He first learned about bitcoin one day in February 2011, from his father, who had a small software start-up of his own. He was seventeen at the time and had recently quit playing *World of Warcraft* for hours on end. Perhaps he was looking for his next big fix. Maybe it was the cryptographic algorithms that appealed to him. One thing was clear, Buterin wanted to get his hands on some bitcoins. There was only one problem: he neither had the cash to buy them nor the computing power to mine them. So he figured out another way to earn them. He started writing for a blog, *Bitcoin Weekly*, where he was paid five bitcoins per post, worth around $4 apiece at the time. In September 2011, he co-founded his own magazine, *Bitcoin Magazine*, with a Romanian programmer called Mihai Alisie.[9] 'The industrial revolution allowed us, for the first time, to start replacing human labour with machines,' Buterin wrote in the magazine. 'But this is only automating the bottom; removing the need for rank and file manual labourers . . . Can we remove the management from the equation instead?'[10]

Through his writings and conversations with early enthusiasts, Buterin realized that bitcoin's underlying blockchain technology was going to be a much bigger deal than the currency itself. That it represented more than just a way of tracking money. He believed it could be used as a powerful tool to re-architect financial, social and even political systems all over the world.

And so at the age of nineteen, in the typical pattern of entrepreneurial tales, Buterin dropped out of college. Using a pile of bitcoins he had earned and that had now significantly appreciated in value, he went round the world. He travelled from bitcoin conferences in San Francisco and Los Angeles to meet-ups in Israel, Amsterdam, London, Barcelona and dozens of other places. He spoke extensively to programmers and dabbled in a few coding projects here and there. All the time, he was trying to figure out how best to make a significant contribution to this growing quasi-cyber-religion.

Huddled around laptops with other early enthusiasts, Buterin would raise question after question. How could they make the system inclusive and open to everyone? How could they create a new economy in which anyone could participate on their own terms? Could companies be run by autonomous algorithms instead of directors? He was not on a technological crusade so much as a mission to upend current power structures. And the rallying cry was *decentralization*, a situation where users, not governments, banks or big companies are in control. 'Ultimately, power is a zero sum game,' Buterin says, 'and if you talk about empowering the little guy, as much as you want to couch it in flowery terminology that makes it sound fluffy and good, you are necessarily disempowering the big guy. And personally I say screw the big guy. They have enough money already.'[11]

He didn't just want to disrupt the big financial institutions. Imagine a marketplace where people can buy and sell anything from artwork to books to honey directly to each other without intermediaries. In other words, imagine Amazon without Amazon (the company, the middlemen and the fees). Similarly, imagine Uber without Uber, a network where drivers could directly offer rides to passengers.[12] Buterin wanted to create a technology that could redistribute power, away from the rent seekers and incumbent middlemen and back into the hands of the people creating value. His vision was the ultimate techno-libertarian promise of creating decentralized marketplaces that nobody owns.

Following the siren call of Satoshi, Buterin released a white paper in November 2013, outlining the plans for his new technology called Ethereum and a currency called ether.[13] The paper is full of technical jargon outlining the problems of the original blockchain, including the key issue of scalability. The blockchain Satoshi created has a built-in hard cap of one megabyte, or about 1,400 transactions per block, that is processed and added to the blockchain roughly every ten minutes.[14] This works out around three to seven transactions per second. To put that in context, Visa handles more than 1,700 transactions per second in America alone.[15] In other words, the bitcoin blockchain is not fast or big enough to handle large volumes of transactions.

The other problem Buterin identified is that Satoshi deliberately designed the original programming language to limit what the blockchain could do – its job was only to store and transfer value. I can send you a bitcoin, you can send it to me. It was not intended as a general software platform on top of which other applications could be created. Imagine the original blockchain as a pocket calculator that can only do a set number of things; Buterin wanted to create the equivalent of a smartphone.

Indeed, the young entrepreneur is sometimes hailed as the next Steve Jobs, although he himself prefers comparisons to Linus

Torvalds, the creator of Linux software. Either way, Buterin is most definitely a 'crazy one': a round peg in the square hole, not fond of rules and with no respect for the status quo, as the famous Apple ad goes. It seems, however, that Buterin isn't trying to create a multi-billion dollar company that will bring him a windfall IPO payday. Ethereum is currently set up as a non-profit foundation based in Zug, Switzerland. And Buterin is its chief scientist.

His vision is to build 'the Lego of cryptographic finance', to give people the building blocks to create all kinds of digital services right out of the box, such as Transactive Grid, a distributed energy market that enables people to buy and sell energy directly from each other.[16] There's also the likes of Ujo Music, which is working to create a platform for musicians to register digital rights, and to be paid directly, without labels, iTunes and other middlemen. Grammy award-winning singer-songwriter Imogen Heap was the first artist to release her single, 'Tiny Human', on the Ethereum blockchain.

Ethereum is based on a stripped-down, Turing-complete programming language known as Solidity. It is simple enough that developers can easily build decentralized apps ('DApps' for short) on top of it. 'Instead of creating a device that just does a specific number of things, you have a device that understands and supports this programming language and whatever people want to do can potentially be implemented,' Buterin explained in an interview with *The Economist*.[17] There are other projects, such as BitCloud, BitAngels and QixCoin, trying to achieve a similar goal. But Buterin, like Jobs, is pretty convincing when he argues that Ethereum is *the* open blockchain worth getting behind.

Ethereum is designed to allow developers to spawn blockchain-based offerings that fall into three main categories or 'buckets'. The first bucket is based on being able to transfer any kind of asset – from shares to concert tickets – in a fast and transparent way on the blockchain. For example, Colu, a Tel Aviv-based start-up founded in 2014, has developed a mechanism to inject every bitcoin with a 'dye' that adds extra data to transactions. Think of it as colouring the cryptocurrency with information about 'real-world' assets, such as the ownership history of a car, which sticks to the coins as they are transferred and stored on the blockchain. Colu creates the equivalent of a digital ID for the asset that can be transferred directly between people, instantly and securely.

The second bucket of applications uses the blockchain to track the supply chain of products, from their provenance to the hands of the customer. Take pharmaceutical drugs. According to Havocscope, which tracks black markets around the world, drugs are *the* most counterfeited products. It is estimated that people pay $200 billion a year for drugs, from Viagra to diet pills to flu medicine, that are not what they say they are.[18] And the consequences of a cancer sufferer taking drugs they think are genuine but are counterfeit duds can be dire. Accenture is currently experimenting with using blockchain technology to create an open and trusted record of where drugs have come from and to track closely what happens to them across the supply chain. It is one example among many of using the blockchain as a kind of *truth machine*.

The DAO fund falls into the third, and perhaps most ambitious, bucket of blockchain applications: *smart contracts*. A smart contract is essentially a digital agreement, whether it's for a loan, job or an investment, which lives on the blockchain. The main difference between it and a traditional contract is that the clauses are not

BLOCKCHAIN: WHAT

1. TRANSFER OF ASSETS — **2. SUPPLY CHAIN CERTIFICATION** — **3. SMART CONTRACTS**

written in English and executed by lawyers. Instead, smart contracts are written in code, with preprogrammed clauses that automatically execute themselves following a set of instructions that work on a principle of 'If this happens, then do this. If that happens, then do that . . .' In other words, the contract is self-fulfilling and carries out what it has been coded to do. For example, my will could be turned into a smart contract, with the rules of how assets should be transferred enshrined in code. Family power plots, squabbles and lies over who gets what would be a battle with, well, a computer program. It's not currently legal but it could be one day. (Even so, it won't replace lawyers altogether. Say the language of my insurance policy was unclear, it would still need a qualified human to make a judgement.)

Self-executing smart contracts need to know there will be a clear outcome. Take gambling. From a young age, I learned about odds and probabilities through betting. (I know, it's a unique approach to teaching your child mathematics.) My family are by no means gamblers but it's a bit of a tradition to back a horse in the Grand National or to pick which side will win the football league. I have such fond memories of sitting in our lounge, huddled around the television, cheering my horse to win and willing my dad's to fall at one of the fences. It made for some wonderful family banter. Now, say I place a bet with my dad on who will win the Wimbledon men's tennis tournament. I pick Rafael Nadal; he picks Roger Federer. We agree on odds and assign a certain amount of ether to a smart contract. On the day of the match final, the system would check the final score of the game via the web and distribute the funds to whoever placed the right bet. We would not need to rely on a bookie. But as the DAO theft demonstrated, smart contracts are only as good as the people who program them: the code will always be susceptible to human error and/or avarice.

On 17 June 2016, an anonymous hacker (or group) exploited a loophole in the DAO fund smart contract. Essentially, there was a programming mistake in the code that allowed a DAO shareholder to create an identical clone fund (known as a 'child DAO') and then freely to move money. And that is exactly what the hacker did: the equivalent of approximately $60 million in ether was drained out of the original fund into the clone.[19] 'I have carefully examined the code of the DAO and decided to participate after finding the feature where splitting is rewarded with additional ether,' the hacker wrote in an open letter explaining the loophole. 'I have made use of this feature and have rightfully claimed 3,641,694 ether and would like to thank the DAO for this reward . . .'[20]

'Rightfully claimed' is the key phrase. It wasn't fraud. Blockchain purists and even some lawyers argued that the hacker was rightfully entitled to the stolen riches. 'I am disappointed by those who are characterizing the use of this intentional feature as "theft",' the hacker added in his letter. He wasn't a Madoff character, duping people. It was clearly the fault of the code in the smart contract that ran on the Ethereum blockchain.

When it became clear that nearly a third of the DAO's funds had disappeared, people on online forums were calling out for one person in particular: 'Where is Vitalik?' one person asked, 'Vitalik, our alien overlord, please save us.' Buterin happened to be in China at the time, figuring out with others in the Ethereum Foundation how to proceed.[21] 'DAO token holders and Ethereum users should sit tight and remain calm,' Buterin wrote on the Ethereum Foundation's blog after the attack.

A very contentious ethical debate followed the heist. Should the community respect the rule of the smart contract and accept the unfortunate consequences? Or should they figure out a way to retrieve the 'stolen' funds?

The attack couldn't be reversed but there was another rule programmed into the contract that would provide a possible remedy. The code imposed a waiting period of twenty-seven days before any money could be paid out of a new fund. So the attacker couldn't do anything with the $50 million for almost a month. 'It's like stealing the *Mona Lisa*,' says Stephan Tual, the COO of Slock.it. 'Great, congratulations, but what do you do with it? You can't sell it, it's too big to be sold.'[22]

The Ethereum team, including Buterin, proposed something called a *hard fork*. It is technical jargon for essentially rewriting history or changing the rules. It is, ultimately, a last-resort solution. Buterin proposed creating an entirely separate version of the ledger that wouldn't have the original loophole. 'It's a one-time fix to a one-time problem,' Buterin said. But first he had to convince the majority of people in the Ethereum network – more than 51 per cent – it was the right way forward.

Supporters of the hard fork insisted that even though they were in uncharted legal waters, they should turn to traditional English law: a contract should be interpreted by the intended spirit of those who wrote it, not the literal interpretation of the words ('to the letter'). Did the intent behind the DAO contract trump the hundreds of lines of computer code?

Those who strongly opposed the hard fork argued that it was a blockchain sin, against the mission of Ethereum, which is to be a 'decentralized platform for applications that run exactly as programmed without any chance of fraud, censorship or third-party interference'. Isn't the goal of a decentralized network that no one has the power to rewrite history, or else the network itself becomes untrustworthy?

The hacker asserted that smart contracts are their own arbiters: 'A soft or hard fork would amount to seizure of my legitimate and rightful ether, claimed legally through the terms of a smart contract. Such fork would permanently and irrevocably ruin all confidence in not only Ethereum but also in the field of smart contracts and blockchain technology.'

When it was put to a vote, 87 per cent of the Ethereum network said 'yes' to a hard fork.[23] The result? The transactions were effectively made void and the millions of 'stolen' ether tokens were retrieved and returned to the DAO crowd investors. It was like the hack never happened. But at what cost?

Buterin dismissed the hack as a 'rite of passage' for a technology still in its infancy. Other DAO code writers and creators, like Christoph Jentzsch, took a similar view. It was a young concept, he argued, and this DAO, with its massive and rapid crowdfunding, had been forced to run while it was still getting the hang of walking. That may be true, but the fork set a dangerous precedent for Ethereum and its quest to become the trusted operating system of the future.

If certain people can reverse transactions, doesn't that mean *they*, not code, are in charge of the system? And if you bend or change the rules once, what happens next time it fails or doesn't suit you? It's rather like the government bailing out the banks when risky trades went south. The hard fork was a top-down reordering of events.[24]

'Its creators hoped to prove you can build a more democratic financial institution, one without centralized control or human fallibility,' writes Klint Finley in an article in *WIRED*. 'Instead, the DAO led to a heist that raises philosophical questions about the viability of such systems. Code was supposed to eliminate the need to trust humans. But humans, it turns out, are tough to take out of the equation.'[25] In other words, even if the maths works perfectly, trust is not simply a matter of code. At the end of the day, the problems of the DAO fund are not just technological but people problems. Humans get in the way.

Andreas Antonopoulos, author of *Mastering Bitcoin*, calls the blockchain 'trust-by-computation'.[26] Reid Hoffman, venture capitalist and LinkedIn founder, labels it 'trustless trust'.[27] But these terms are a bit misleading – there is still clearly trust involved. You have to trust the idea of the blockchain; you have to trust the system. And given that most people lack the technical know-how to understand how the system really works, you have to trust the programmers, miners, entrepreneurs and experts who establish and maintain the cryptographic protocols. A large dose of faith is required. But it's true to say that you don't have to trust another human being in the traditional sense.

So here, with the DAO fund and the hard fork, was a trust stumble. History, however, is littered with high-minded projects that were pushed through – by kings, emperors, inventors, scientists, surgeons – before they were completely ready. Early steel bridges that collapsed, experimental operations that killed the patient, explosions in the lab, re-engineered ships that sank. The short-term results were disastrous but valuable lessons were learned. As polymath Danny Hillis said back in 1997, 'Technology is everything that doesn't work yet.'[28]

Every innovator wants to be first over the line, and it's no different with the quest for the ultimate blockchain technology. Inevitably, there will be glitches along the way because that's how innovation comes into being and grows resilient, just as the body develops its immune system by being exposed to bugs and viruses. The blockchain's enormous potential means developers and investors are taking a classic 'fail fast, fail forward' approach. As Ethereum's story shows, even the odd hiccup and regrettable repair job won't stop people jumping on the blockchain juggernaut.

Since 30 July 2015, the day Ethereum went live, around 1,211 DApps have been created on its blockchain.[29] To put that figure in context: when the Apple App Store launched in 2008 there were 500 apps available. By 2010, there were 250,000 and in June 2018 there were 2.1 million.[30] Investment funding in blockchain-related start-ups has increased globally from an estimated $1.3 million in 2012 to more than $1.4 billion in 2016, according to PwC.[31] Much of the interest (and a lot of hype) is focused around how decentralized ledgers can create a shared version of *single proof* or a *digital truth* about the identity of assets. 'I see what you see . . . and I *know* that what I see is what you see.'[32]

Take diamonds, for instance. The round diamond in my engagement ring has an interesting history. When my family, Eastern European Jews, fled Russia in the late 1800s they exchanged their wealth for three or four diamonds. Precious stones were easier than money to hide, and were commonly sewn into the linings of coats or hidden in the soles of shoes. My nana, the late Evelyn Amdur, supposedly had the last remaining family diamond, roughly 2.3 carats, in her possession. She wanted me to have it when I got married, so she asked my then fiancé and now husband, Chris, if he would use it in my engagement ring. He had it designed in a beautiful antique setting. It's a stunning ring. A few years ago, however, the stone started chipping quite badly, with chips you could see and feel. The stone was meant to be valuable, without inclusions, so it was an odd thing to happen. I started to suspect the stone was possibly not the original family diamond.

Evelyn was like a character in a *Catherine Tate Show* sketch – funny, frank, slightly crude and with a remarkable talent for fooling people. A couple of years ago, she was sick and it was clear she didn't have long to live. One day when I went to visit her, I decided to ask her about the ring. She was sitting comfortably in her favourite beige armchair, cup of tea and digestive biscuit in hand, when I raised the sensitive topic. Nana smiled, with a twinkle in her eye, and said, 'Well, darling, the luxury cruise was very wonderful.' I never did get a straight answer before she passed away.

Sure, I can go to a jeweller and get an assessment of the diamond's four Cs – cut, clarity, carat and colour – and know its real value. But that's not what is important to me. And I do kind of like the idea that my nana, the rascal, sold the family heirloom and went on the trip of her dreams. Even so, I would love to have known the life story of the real stone – its age, its lineage and where it is now. The problem is that the information, like many other valuable items, was stored on a paper record or certificate that has been lost years ago. It is the sort of thing that happens all the time. The blockchain, however, offers a way to capture and keep the history of an item – whether it's a diamond, a valuable stamp, bottle of wine or piece of art.

Thanks to consumer pressure, we can now find out the source of, say, fair trade coffee from Starbucks or the organic cotton in Gap T-shirts, yet we still know surprisingly little about most of the items we own and use. Was that organic, grass-fed cow really raised on such-and-such free-range farm as claimed, slaughtered at such-and-such abattoir, packaged last week and brought to the supermarket on Wednesday? Or, as in the Tesco scandal, was it contaminated with some horsemeat at some point in the journey from pasture to plate? Is this product what it claims to be? Supply chains and the origins of a product are, for the most part, a dark secret.

Provenance. It's a word Leanne Kemp, a serial entrepreneur, has spent a lot of time thinking about. 'It means the history of something, where it came from and where did it go,' she says in her distinct Aussie accent.[33] 'Who owns it? Who sold it and where is it now?' Essentially, it's the life story of an item, and in the world of goods, especially expensive or rare items, provenance matters. The Yapese were right with their fei stones – the value of an item should not be separated from its origin and history.

Kemp, now in her late forties, was born and raised in Brisbane, Australia. She moved to London in the late 1990s and divides her time flying between continents, her work specializing in an unusual blend of technology and the jewellery trade. 'I'm a technologist. A "super-nerd" who can cut code,' she says. 'I used to work in RFID [Radio Frequency Identification] to track the identity of goods as they moved through the supply chain.' Kemp likes to use a new technology to solve a problem in a way it hasn't been solved before.

Several years ago, she began to immerse herself in the world of cryptocurrency. From the outset, like Buterin, she was far more interested in the ledger technology than the bitcoin itself. 'The currency had been on the market for quite some time. The fundamental change occurred in 2014 when the bitcoin network released something called an *op return function*,' she says. 'Basically, it enables you to trade a coin and to hash on to the coin a piece of data. I started to think about what I could do with that functionality.' Could it be used to track the origins and ownership of, well, anything worth tracking?

Looking into the diamond industry, Kemp discovered that it was plagued by problems like synthetic diamonds, insurance fraud, theft and tampering of paper certificates. Some £45 billion is lost annually in the United States and Europe on insurance fraud alone, and an estimated 65 per cent of fraudulent claims go undetected.[34] Then there are the notorious 'blood diamonds', precious stones mined in African war zones, often by young children, with the funds from sales frequently used to arm brutal rebel conflicts. By the time Kemp came along, there was at least one certification system, the Kimberley Process, in place to reassure buyers they weren't buying a diamond with blood on it, but that system was not foolproof and still largely depended on paper trails.

Early in 2015, Kemp sketched the idea for her company on the back of a napkin. She drew the trail of a diamond from a mine to a marketplace to a person's finger. 'I sketched the data we could store against the provenance chain of diamonds and that's where it all started.' A few months later, she founded Everledger, a London-based start-up that digitally certifies diamonds on the blockchain.

'We create a diamond's digital thumbprint or ID,' Kemp says. 'Take a three-carat diamond. It will have a serial number inscribed on its girdle during the grading process. There are the four Cs – cut, clarity, carat and colour – but there are forty other attributes, such as angles, cuts and pavilions, which make up that specific diamond.' The diamond's ID is enshrined in the blockchain, creating an immutable record for insurers, traders and customers to know the real provenance and movements of a diamond over its entire history. 'We can apply this technology to solve very big ethical supply chain problems: from ivory poaching to blood diamonds,' says Kemp, adding that because a diamond moves through so many hands, corruption, scams and rip-offs can happen at any stage of its journey from grimy mine to gleaming boutique. 'Blockchain technologies allow us to bring ethical transparency on a global scale.'

So far, Everledger has digitized the ID of more than 1 million diamonds and has partnered with big financial players including Barclays and Lloyds. The company is also building an anti-counterfeit database. What this means is when a stolen diamond resurfaces for sale on an online marketplace such as Amazon or eBay, it will be much easier for investigators to track its history and return it to its rightful owner.

In a few years, we will get to a point where we are able to check the provenance of all kinds of items before we buy them and find out precisely if they are what they claim. The question is, will customers want to know and will they care? 'We've seen this with organic food, but when it comes to luxury goods, people mostly rely on the trust of a brand such as De Beers to decide what to buy,' Kemp says. 'If there was to be a full consciousness of transactions, what would we choose? I don't know the answer to that yet.'

Everledger is essentially building a platform to track the true identity and reputation of objects. 'We're on the next generation of technology where transactional data of objects and assets becomes woven into the web and that's what I think is the World Wide Ledger,' says Kemp. The World Wide Ledger (WWL) offers a way to make and preserve truths – the history of assets.

Imagine if this technology had been around in the Second World War during the greatest art theft in history. Art objects of all kinds, deemed suitable to Hitler's taste, were shipped in freight cars from all over Europe to end up in Germany. It is estimated that more than 650,000 artworks, including Giovanni Bellini's famous *Madonna and Child* and one of Edgar Degas's iconic ballerina paintings, were looted by the Nazis, with many of the works shamelessly stripped from the homes of wealthy Jews who had fled or been sent to concentration camps.

When the war was over, the Allies put together a special unit of personnel, the so called 'Monuments Men', devoted to finding and returning looted art to their rightful owners. Despite those efforts, the art was largely returned to countries, not to individuals. Some 100,000 stolen works of art remain unaccounted for, their current location and owners an enigma.[35]

It turns out that during the 1950s and 1960s, hundreds of works were sold, at a significant discount, to the Nazis who had stolen them. Perhaps more alarming was the fact that Jews were made to buy back in auctions (or at least split the fees with houses) works they had proved to be rightfully owned by their families. 'They called them a "return sale",' says Anne Webber, the founder of the Commission for Looted Art in Europe, a London-based non-profit.[36]

Naturally, museums from the Tate in London to the Metropolitan Museum of Art in New York and the Louvre in Paris try to check the ownership history of works on display and in storage, but it is an incredibly complex task. In most instances, the museum has no way of verifying the claims of ownership of plundered property because, just like my diamond, the paper records have either been lost or tampered with along the way. Remarkably, it wasn't until 2001 that the American Alliance of Museums published the first set of strict guidelines for handling and checking the provenance of *Raubkunst*, Nazi-confiscated art.[37]

So how do families, many of whom lost parents in the Holocaust, even know what is missing from their collections? How do you stop people filing fake claims if the true existence of the artwork is unknown? Decades after the war, most attempts at recovering and returning looted art fail, and the knowledge about collections is disappearing as the original owners pass away.

One problem with *Raubkunst* has been that the art world, like many other industries, lacks transparency. In May 2016, Everledger announced an investment and partnership with Vastari, a company that creates a network between museums and collectors. 'There are a number of individuals around the world who hold significant collections, say, in Andy Warhols,' says Kemp. 'If they could easily track the movement of their artwork, they might be able to realize or release the financial potential of that piece and the public would see it more.'

Kemp is just one of many entrepreneurs around the world recognizing the power of the blockchain to act as a new kind of digital trust broker. And what works for diamonds could also work for, well, fish. Provenance, founded by Jessi Baker, is using the technology to track the supply chain of fish from fisherman to plate. 'Every product has a story,' says Baker.[38] 'There is often an enormous gap between advertising and the reality of operations, what goes on behind the scenes. I find it strange that the provenance and production of products has remained so secret for so long.'

It's not just start-ups thinking this way. Alibaba wants to weed out counterfeit foods – whether soy sauce made using dirty tap water or fake spices unfit for human consumption – by using the blockchain to track products sold on Taobao and Tmall through the supply chain. Retail giant Walmart and IBM have partnered with Tsinghua University in Beijing digitally to track the source, factory information and movement of pork in China on a blockchain. 'Consumers today want more transparency about where and how a product came to be,' says Frank Yiannas, vice president of food safety at Walmart. 'If you shine a light on the food system, that leads to transparency.'[39]

The transparency Yiannas is referring to holds immense value in industries where lies and falsehoods are rife. Indeed, at the World Economic Forum at Davos in 2017 the agenda was full of presentations with titles such as 'Blockchain Revolutionizes Global Transactions' and 'Employing the Blockchain to Serve Society'.[40] One of the much-hyped virtues of blockchain technologies is their potential for emerging markets such as Honduras and Ghana where weak governance and a lack of record-keeping systems means that trust is all too often in short supply. 'It has the potential to leapfrog billions of people into a new era – in parallel to the way that mobile phones helped them leapfrog over landlines,' says Mariana Dahan, a senior operations officer at the World Bank.[41]

Hernando de Soto Polar, a prominent Peruvian economist, has long held a view that capitalism will only thrive in the Third World when people feel that the law is firmly on their side, or even simply applies to them and their personal circumstances. 'What you have to remember is that people outside the legal system are the majority. There are 7 billion people in the world, and those who are outside the legal system are five billion,' he says. 'So this is no marginal phenomenon.'[42] One of de Soto's core arguments is that a lack of clear *de facto* property rights, not capital, is what has held (typically poor) people back in developing countries for so long.[43] An estimated 5 billion people, mostly in the developing world, have difficulty proving they own land, businesses or cars.[44] With no legal owners, the assets are effectively walled off from the 'official' economy. It adds up to trillions of dollars in locked or 'dead capital'. If ownership of land and housing is clear, recognized and protected, people will look after what they can control. It also gives people collateral to borrow against.

In Ghana, where it is estimated around 78 per cent of land is unregistered, homeowners will often put a sign outside their house painted with the words 'This house is not for sale'.[45] It is often the only means available to show that a place is occupied. Another big problem that arises from the unregistered land is that people sell what they don't rightfully own. Government officials, abusing their power, sometimes buy an election vote here and there using fraudulent land titles. Bureaucrats sometimes hack databases to get themselves the rights to a nice beachfront property. So start-ups such as Bitfury, ChromaWay and Bitland are starting to work with governments to collect and organize property records on blockchains, based on the premise that every *coin* on the blockchain could represent a unique house or land title. If there were an immutable record of land-title information, theoretically at least, dodgy officials would not be able to tamper with the record without leaving a digital trail.

The UK government's chief scientific advisor, Sir Mark Walport, published a report in January 2016, claiming that 'distributed ledger technologies have the potential to help governments collect taxes, deliver benefits, issue passports, record land registries, assure the supply chain of goods and generally ensure the integrity of government records and services'.[46] Similarly, the Dubai government announced its ambitious plans to go paperless and move all documents on to the blockchain by 2020.

For the moment, the blockchain remains in the realm of enthusiasts, innovators and idealists. It's still unclear what will be the killer consumer app that takes it into the mainstream. But one thing is certain: the intermediaries and centralized behemoths such as banks and accountancy firms will do their darnedest not to be cut out of the picture by a network of digital ledgers. Indeed, the first industry widely to adopt the blockchain could be the very middlemen Satoshi wanted to replace: finance.

The first sentence of Satoshi's 2008 bitcoin paper defines it as follows: 'A purely peer-to-peer version of electronic cash that would allow online payments to be sent directly from one party to another without going through a financial institution.'[47] 'Without a financial institution' is the key point. But it looks like banking middlemen are going to use the technology to make the exchange of money faster and cheaper. 'My hunch is that the blockchain will be to banking, law and accountancy as the internet was to media, commerce and advertising,' says Joi Ito, the respected entrepreneur, professor and director of the MIT Media Lab. 'It will lower costs, disintermediate many layers of business and reduce friction. As we know, one person's friction is another person's revenue.'[48]

So if financial incumbents can't be at the centre, why not control the new trust architecture that transactions will flow through? Indeed, patent wars are brewing and the race is on to try to 'own' blockchain technologies.

Anyone on Wall Street knows who Blythe Masters is. She is a banker's banker. Born in 1969, she was raised in southeast England, where she attended the exclusive King's School in Canterbury. Her accent is cut-glass English: proper Home Counties. During her gap year, before studying economics at the University of Cambridge, she interned at JP Morgan Chase. At the age of twenty-two, she officially joined the bank in the derivatives team in New York. Her rise through the ranks was inexorable. By the time she was twenty-eight, she was managing director, the youngest woman ever to achieve the title in the investment bank's long history. At thirty, she became head of the global derivatives unit. At thirty-four, she became chief financial officer, joining an elite group the media dubbed as the 'JP Morgan Mafia'. She became known on Wall Street as 'Queen of Commodities'.[49]

In 1989 more than 10.8 million gallons of crude oil from the oil tanker *Exxon Valdez* spilled into the pristine oceans of Alaska, spreading far and wide. The potential damages were estimated to be upwards of $5 billion. The oil company needed a loan, an enormous one. So in 1994 they went to their long-term bank, JP Morgan. Masters happened to be leading the team managing the financial side of the crisis for the oil company. Exxon were an old client and the bank didn't want to turn down their request. At the same time, the loan was risky and would tie up a lot of the bank's reserve cash. Masters came up with what seemed at the time like an ingenious idea: what if the risk of the loans could be sold?

Her thinking was based on the fact that investment banks already swapped bonds and interest rates. So why not swap the risk of defaulting on loans? And so the idea of the 'credit default swap' was born and took off big time. Blythe Masters was credited as the mastermind behind the concept. Credit default swaps were meant to reassure investors the risk was hedged if a loan went south. Instead, as we now famously know, it blew enormous holes in the balance sheets of banks and insurance companies such as American International Group (AIG) and mortgage lenders like Fannie Mae, who didn't have the collateral owed when the underlying credit swaps deteriorated.

During the 2007–2008 financial crisis, Masters had the courage, or some might say audacity, vociferously to defend the bank's trading activity. She asserted they had done nothing wrong, despite the millions of Americans who lost their homes and jobs as a result of the crisis. If the media needed a target, Blythe Masters had a giant circle on her back. Warren Buffett went so far as to describe the derivatives she masterminded as 'financial weapons of mass destruction'. She was vilified, some say unfairly, given how many others jumped on board with her and then ran for cover. In April 2014, after almost three decades with JP Morgan, Masters resigned.

Now, she is back, championing not swaps but another potential money-spinner – blockchains. 'You should be taking this technology as seriously as you should have been taking the development of the internet in the early 1990s,' Masters told a packed and rapt audience of money managers and investors at a conference held at the Le Parker Meridien Hotel in mid-Manhattan in the summer of 2015. 'It's analogous to email for money.'[50]

Around the time when bitcoin and blockchains were starting to catch the attention of the mainstream investment world, a New York-based start-up called Digital Asset Holdings (DAH) was launched. Blythe Masters was at its helm. The Wall Street veteran is knowledgeable about a common problem many banks face – getting incompatible financial databases to talk to each other. It's costly, complex and takes time. While it might seem that traders work at Red Bull speed in lightning-paced environments, the technology used to execute trades is remarkably old-fashioned and *slow.* Lots of phone calls are made, emails traded and even the occasional fax is still sent. It can take up to three days – T3 – for stock trades to change hands via clearing houses such as the National Securities Clearing Corporation (NSCC). It's a process known as 'settlement lag'. Every hour before settlement happens, when a trade precariously hangs between sale and purchase, increases the risk that the trade won't go through. Obviously, it's in the banks' interest to close that lag time as much as possible.

Blockchains could help reduce the gap of the entire lifecycle of a trade from days to minutes, even to zero. According to a report by Santander InnoVentures, the Spanish bank's fintech investment fund, by 2022 ledger technologies could save banks $15–20 billion a year by reducing regulatory, settlement and cross-border costs.[51]

Digital Asset Holdings wants to be *the* distributed database handling these speedy transactions. And the who's who of the world's biggest financial names, including Goldman Sachs, Citibank and Blythe Masters's old employer, JP Morgan, have ploughed more than $60 million of investment into DAH. Speed and efficiency are not the only qualities that make distributed ledgers attractive to banks. 'Regulators will like that blockchain-based transactions can achieve greater transparency and traceability – an "immutable audit trail",' Masters says.[52] In other words, it could help eliminate the kinds of fraud that come from cooking the books. It's rather ironic that these words come from a woman who spent several months being investigated by the Federal Energy Regulatory Commission for a cover-up of energy-trading strategies. Masters was not cited for any wrongdoing and no action was brought individually against her. JP Morgan paid $410 million to settle and close the case, without denying or admitting wrongdoing.[53]

On Wall Street, the race is on to embrace or control what could be either its biggest ally or its death knell. Where does the average Joe store their money? In a bank's current or savings account or a safety deposit. But the blockchain could become a new repository of value.

How do typical loans work? A bank assesses the credit score of an individual or business and decides whether to lend money. The blockchain could become the source to check the creditworthiness of any potential borrower, thereby facilitating more and more peer-to-peer financing. How do typical credit cards and money transfer services work? They currently flow through a bank, but the blockchain could handle this exchange of value directly from person to person. Consider traditional accounting, a multi-billion industry largely dominated by the 'big four' audit firms, Deloitte, KPMG, Ernst & Young and PwC. The digital distributed ledger could transparently report the financial transactions of an organization in real time, reducing the need for traditional accounting practices. And that is why most major players in the financial industry are busy investing significant resources into blockchain solutions. They have to embrace this new paradigm to ensure it works for, not against, them.[54]

A San Francisco-based venture called Chain is said to have raised more than $30 million in funding from big names such as Nasdaq, Visa and Citi Ventures to develop open-source code for a distributed ledger. IBM, Wells Fargo, the London Stock Exchange and others have joined forces with Digital Asset Holdings to develop blockchain software that is also open source, making the underlying recipe available to developers. Originally dubbed the Open Ledger Project (and later renamed Hyperledger), the joint efforts are being overseen by the widely respected Linux Foundation.

Goldman Sachs has recently filed a patent for its own cryptocurrency, its own version of bitcoin, called SETLcoin which processes foreign-exchange transactions. It is designed to run on the bank's own private blockchain. This means the replicated ledger of transactions still sits behind the closed walls of the bank, centralized and guarded. It seems to defeat the very purpose of the technology, which is to create a single indisputable version of the truth, freely accessible to all, that could eliminate the need for the bank entirely. In the patent, Goldman describes SETLcoin as having the potential to guarantee 'nearly instantaneous execution and settlement' for trades.[55] It would mean all the capital the bank is required to keep in reserve, to hedge against the risk of transactions if they don't settle, would be freed up.

More than forty banks have a stake in a consortium called R3CEV to come up with shared standards for blockchains.[56] The technology will be pretty much worthless if there are multiple versions of the blockchain that can't work together. R3CEV wants to bring along all the banks and regulators so they can share just one – a ledger that is not controlled by any one person or organization but by many participants. Sure, it's collaboration, but perhaps not the kind Satoshi had in mind.

Notably, R3CEV has recruited a man by the name of Mike Hearn as its chief platform officer. The former Googler is a big deal in the blockchain world. Hearn spent more than five years working full-time alongside Gavin Andresen, as part of Bitcoin Core, the original group of developers that maintain the open-source code that runs the bitcoin peer-to-peer network.[57]

Hearn admits he is a 'tell-it-like-I-see-it kinda guy'. In January 2016, he publicly denounced the future of bitcoin and said it was inherently doomed. 'It has failed because the community has failed. What was meant to be a new, decentralized form of money . . . has become something even worse: a system completely controlled by just a handful of people,' Hearn wrote. 'The mechanisms that should have prevented this outcome have broken down, and as a result there's no longer much reason to think bitcoin can actually be better than the existing financial system.'[58]

Just days after he published the post, Hearn joined the R3CEV banking consortium. 'The current Bitcoin system, I mean the system we actually use today with the blockchain, isn't going to change the world at all due to the 1mb limit [the maximum size of a bitcoin block],' he said in defence of his move. 'So if I have a choice between helping the existing financial system build something better than what they have today that resembles Bitcoin, or helping the Bitcoin community build something worse than what they have today that resembles banking, then I may as well go where the users are and work with the banks.'

From Buterin to Hearn, it seems that everyone, however different their motives, is in a race to create something like the original Satoshi blockchain, only better. For many, it's the biggest game in town.

The blockchain raises a key human question: how much should we pay to trust one another? In the past year, I've paid my bank interest and fees, some hidden, to verify accounts and balances so that I could make payments to strangers. I've spent thousands of dollars on lawyers to draw up contracts because I am not quite sure how another person will behave (and to sort out a few incidents where trust broke down). I've paid my insurance company to oversee the risk around my health, car, home and even life. I've paid an accountant to reconcile an auditing issue. I've paid an estate agent tens of thousands of dollars essentially to stand between me, the prospective buyer, and the current owner to buy a house. It would seem we pay a lot for people to lord over our lives and double-check what's happening. All these 'trusted intermediaries' are part of the world of institutional trust that is now being deeply questioned.

Many of the ideas surrounding the blockchain sound ambitious, risky and radical. Many are being over-hyped, over-funded and will likely fail. What's not in doubt is that, as the cost of trust plummets because of new technology, the third parties currently paid to facilitate our trust – be they agents, referees, watchdogs or custodians – will increasingly have to prove their value if they don't want to be supplanted by an 'immutable' ledger.

In 1993, enthusiasts such as Al Gore were telling the world about a coming 'information superhighway' that would change the world. The internet was a novel concept few had grasped and people didn't really know what to make of it. John Allen, an early web aficionado, went on TV to try to explain how people would use it: 'In this world, there's a table with a big sign on it that says "Football" and there's 150 or 1,000 jocks all around the world who want to talk about football,' he said on CBC.[59] At that time, Mark Zuckerberg was nine years old. Google was three years from being born. All the other products and companies that would emerge to commercialize the internet and its future potential were not yet clear. Today, it is circa 1993 for blockchain technologies. Even though most people barely know what the blockchain is, a decade or so from now it will be like the internet: we'll wonder how society ever functioned without it. The internet transformed how we share information and connect; the blockchain will transform how we exchange value and whom we can trust.

Conclusion

'It is trust, more than money, that makes the world go round.'

Joseph Stiglitz[1]

On 28 February 2016, in a small cinema in Pan Paper Village in Kenya, more than 400 locals had gathered to watch an English football match. The game was between two of the greatest rivals in the Premier League: Manchester United and Arsenal. The cinema belonged to Eric, an entrepreneur who also owns a printing business, small kiosk and a photo studio. He is a local success story.[2]

The Kenyan crowd, sitting in a darkened room with the game projected against a wall, cheered as the players walked out to start the match. Man U started out strong with Arsenal struggling to get anything going. Then the screen went blank. The crowd let out a groan. 'Get the game back on!', 'Boo, what's happening?', 'We're missing the match!', the paying customers hollered at Eric.

The cable company had disconnected the service because Eric's bill had not been paid on time. Feeling frazzled, with hundreds of eyes trained on him to fix the problem, and fast, he reached for his phone. He opened an app called Tala, a company that makes loans to people without a traditional credit history in emerging markets including Kenya, the Philippines and Tanzania. Luckily, he recently reached gold status for repaying four loans on time and his credit had been increased to 5,000 Kenyan shillings (KES), the equivalent of around £40. It's a decent amount of money when you consider that a small 0.33-litre bottle of Coke costs around 50 KES.

With a few taps and swipes on his phone, Eric took out his Tala credit line and immediately paid his cable fees. Within minutes, the match was back on the big screen. The crowd cheered; the score was still 0–0. Each customer paid Eric around 15 KES, so he made 6,000 shillings in takings. After everyone had left, he paid back his Tala loan and kept the profits.

Eric is one of more than 300,000 people Tala has issued loans to in Kenya since the company was founded in 2011. Loans range from $10 to $100, with terms of three to four weeks. The interest rates range from 11 to 15 per cent, considerably better than those of the loan sharks who charge a crippling 300 per cent and more. The repayment rate is 90 per cent. Tala's customers are part of the one-third of the world's population that is 'unbanked'. Across the globe, there are 2.5 billion people who have no traditional credit score and are therefore not eligible to receive loans to fund new businesses, buy a home and generally improve their lives. 'For a traditional bank, [the lack of a credit score] means there is no real data to answer the question: "What's my basis for investing in this person?",' says Shivani Siroya, the thirty-four-year-old founder and CEO of Tala.

Siroya was first raised in Udaipur, India. Her mother, by all accounts, was a renegade. She went to medical school and was the first female doctor, a gynaecologist, in her community. She ran medical camps in rural areas, training women how to deliver babies more safely. Her relationship with her patients was always selfless. 'We would say to her, "Mom, you work so hard and your patients never pay you,"' Siroya tells me.[3] 'But money wasn't the thing that kept her going. For her, being a doctor is not a transactional relationship. She truly wants to help.'

The family eventually ended up moving to the United States, settling in Brooklyn, and her mother continued her good work there, often allowing poor patients to pay when they could or not at all. 'She gave them credit because she trusted them,' Siroya says.[4]

On the phone from Santa Monica, California, where she now lives, Siroya sounds so calm and tranquil she might have just finished a yoga class. She gives the impression of being, like her mother, full of life and goodness. It's odd to imagine her working in the cut-throat culture of mergers and acquisitions at Citigroup, which is where she began her career as a banking analyst before going on to become an equity researcher at UBS. Siroya loved the data but something didn't sit right with her. 'I didn't want to be analysing companies selling diet drugs; I wanted to work with companies doing something with impact.'[5] She heard about social enterprise guru Muhammad Yunus, who won the Nobel Peace Prize in 2006 for pioneering the concepts of microfinance and microcredit. Shortly after, she quit her job and joined the United Nations Population Fund.

At the UN, she ended up working with thousands of microbusiness owners, such as locals running food stalls and small kiosks across Africa and India. Over the course of two and a half years, she interviewed more than 2,500 people, going door to door and learning the detail of their lives. 'These micro-entrepreneurs were working so hard to grow their businesses and to help create jobs, yet they felt really stuck. They were getting access to really informal micro-loans, often from loan sharks, but they couldn't get access to any kind of real capital in the form of a business loan,' Siroya says. 'When I looked at the other side, I found that traditional banks felt these individuals, operating in an economy where cash is prevalent, were too risky. I saw a gap in the market.' She realized the problem came down to identity. How could she build financial identities for people who didn't otherwise have any?

Approximately a billion people in the emerging markets have basic smartphones. People are using them in the same way that Westerners do – to text friends, surf the net, run their daily lives and pay for everything from electricity bills to parking fees. The phones also tell a story about their owners. Tala can cull more than 10,000 data points from a phone in less than one minute to gauge a person's ability and willingness to repay loans. 'We look at behavioural things such as what are their current spending habits? Do they have consistency in their income? What other apps do they use?' explains Siroya.

The size of a person's network is a strong trust signal for potential borrowers. Turns out, if our phone calls last more than four minutes, we tend to have stronger relationships, and therefore may be more creditworthy. Similarly, people who communicate with more than fifty-eight different contacts tend to be better borrowers because they have a wider network to depend on. Even how we organize our contacts can be revealing. 'If more than 40 per cent of the entries in a person's contact list have both first and last names, it suggests a customer who is sixteen times more reliable than one with very few contacts listed with first and last names,' explains Siroya. Filling in a first and a last name shows, in a small way, the care and attention we pay to something. Indeed, no single piece of information determines whether someone gets a loan – it's the cumulative points of data that provide a clear picture of a person. 'It's a financial identity that looks more like a person and less like a score,' says Siroya. 'This is data that would not be found on a paper trail or in any formal financial record.'[6] It proves that a person doesn't need a traditional credit score to prove they are trustworthy.

Today, Tala is the fifth most used app in Kenya. Only the Bible app, Facebook, Twitter and WhatsApp are ahead of it. 'Customers don't see us as a traditional banking institution which they have a transactional relationship with. They see us as a personalized financial partner,' Siroya says. Tala illustrates how technology can help find ways round trust bottlenecks to unlock more economic activity. It demonstrates how data and algorithms can prove that there are billions of people like Eric, often overlooked and undervalued, that deserve to be trusted.

Why is Tala so successful? It sounds so simple: start with the person, not the system.

'Free markets can succeed for all if business works with the people, not just sells to them,' says Richard Edelman in response to the 2017 Edelman survey that revealed a 'global implosion of trust'.[7] To get out of the current trust collapse, we need radically to rethink the foundations on which our institutions are built so they are designed to work not just *for* people but *with* people. 'We must rebuild faith in the system citizen by citizen, community by community, where common goals and fairness matter,' writes Edelman. In other words, put people at the centre of everything you do.

The trust test for any organization is this: would people describe it as an 'honest, competent and reliable friend', someone who is there when you need them?

To give a simple example, when Tala wants to send their customers a reminder to repay their loan, they don't constantly bombard them with notifications when those customers clearly don't have any money in their bank. What's the point? Instead, they send borrowers a friendly SMS reminder the minute a deposit hits their account. More than 80 per cent of people pay on their phones as soon as they get that reminder. It shows the power of technology when it contains simple ingredients of humanity such as empathy and fairness.

When I started writing this book, I thought, maybe a little naively, that most of the entrepreneurs, hackers, leaders and innovators I would meet designing ideas based on distributed trust would be like Siroya. And many are – working to reframe deeply held institutional assumptions about power, access and equality. There are figures like Gerard Ryle, head of the ICIJ team of journalists behind the Panama Papers story, who uses digital networks to get people to work together, selflessly and collaboratively. Or entrepreneurs such as Leanne Kemp, founder of Everledger, and Savi Baveja, CEO of Trooly, who want to take on such enormous industries as background checks and the diamond trade, industries with troubled legacies and where trust has been systemically broken in many places. If you add up the stories of founders such as Joe Gebbia from Airbnb, Lynn Perkins from UrbanSitter and Frédéric Mazzella from BlaBlaCar, you start to see a world where new mechanisms for enabling trust leaps can make us comfortable with people, ideas and experiences we may never have otherwise considered. Their companies might look very different from each other but what they have in common is using digital tools to build trust with strangers to connect and collaborate on an unprecedented scale.

On the other hand, there are entrepreneurs who act like 'digital gods', reaping the humungous benefits of platforms and algorithms with immense impact on our lives, but denying responsibility when things go wrong. In some cases, the high walls of institutions are merely being replaced by opaque, controlling algorithms and unpredictable leaders of platforms. Travis Kalanick, for example, the controversial co-founder and former CEO of Uber. Among other instances of bad behaviour, Kalanick sent emails to his PR team after the media reported a woman had been choked by a driver, instructing them to 'make sure these writers don't come away thinking we are responsible even when these things do go bad'.[8] The company and its notorious founder don't do themselves any favours by projecting an aggressive and flawed public image.

In late January 2017, a protest movement with the hashtag #DeleteUber quickly gained traction over two issues, both of which involved the company's connection to President Donald Trump. The former CEO said his participation on an advisory group within the Trump administration had created what he called a 'perception-reality gap between who people think we are, and who we actually are'.[9] But perception is everything.

When trust is lost, a company has to exhibit humility, to be unafraid to give a genuine apology and acknowledge mistakes, and to demonstrate a clear willingness to fix what is wrong. Indeed, in an age of distributed trust, it's much harder to get away with errors and bungled responses, because there are far more people watching. Real incidents, or merely unfounded views that someone feels to be true, circulate at an unprecedented speed. A tweet about Uber's surge pricing might appear in one person's feed in the morning and then have flown around the world on social media by noon, leading to a full-blown protest not long after. The consequences for trust are enormous.

In 1919, the Irish poet W. B. Yeats wrote a poem called 'The Second Coming' to describe the atmosphere of post-war Europe.

> The best lack all conviction, while the worst
> Are full of passionate intensity.[10]

Almost a hundred years on, we are living in a time where perilous trust battles are raging between facts and 'alternative facts', falsehoods and rumours; between open platforms and gated communities; between elites or authorities and 'the people'; between the informed, the misinformed and the credulous. It's fair to assume that for some time to come the prevailing mood will be anti-elite and anti-authority – a feeling that our traditional systems are deeply failing us. Trust in big institutions and the established order will continue to unravel and collapse.

That reaction might be understandable but a blanket hammering of institutional trust – a wholesale rejection of the media, the courts and intelligence services, the truth-defending organizations that underpin any democracy – threatens to create chaos. We obviously need to question institutions and hold them accountable, but if we pull the rug out from under them all, what are we left with? Potentially a dangerous *trust vacuum* that is open to manipulation and being filled with catchy conspiracy theories, comforting biases, unfounded accusations and sleights of hand. A trust free-for-all, in other words. Think about it: when people are told they cannot trust any of the old institutions, they can end up trusting nothing, or anything. Institutions do not need to go the way of the dodo – they just have to learn to adapt to this new trust landscape if they don't want to be left behind.

When institutional systems fail, alternatives will always rise up to take their place. Distributed trust in itself can't knock down the rise of extremist populist movements, dangerous policies introduced by radical political leaders or a divisive resurgence of nationalism. But, driven democratically and rationally, and shaped and reshaped by people's needs and innate preferences about how they want to do things, it can provide a path forward for businesses, governments, media and other key institutions. It gives them a means to redesign systems that put people first in ways that are more *honest*, *inclusive* and *accountable*.

Significantly, this revolution is taking place in a landscape of rapidly shifting and evolving technologies, where the once unthinkable, the once impossible, can become the new normal in the blink of an eye. Humans are naturally attuned to taking trust leaps. Today, however, it feels like we are constant 'newbies' leaping at such an accelerated rate and in so many realms at once that it's dizzying.[11] That's another challenge; setting up trust systems that can adapt and keep pace with an unprecedented rate of change.

I would not have written this book if I did not believe in the enormous potential of distributed trust to give people, even countries, the tools and power to leap out of low-trust situations; if I didn't believe in its ability to help us find ways through the treacherous storm of distrust we are currently only just weathering. There is a large dose of optimism and exciting potential in the world of distributed trust, although it would be foolhardy not to acknowledge there is also a high degree of fear and uncertainty. It's a work in progress. We're still discovering its strengths, its virtues and its vulnerabilities.

Over the course of *Who Can You Trust?* we've looked at several stories where distributed trust always seems to lead us back to centralized power; a take-over, if you like, of those early good intentions. Take Amazon, Alibaba or Facebook. They might have begun as ways to democratize commerce or information, but they have become centralized behemoths in control of valuable and ever-more sensitive data.

What's more, institutions meant to keep dominating powers in check – regulatory bodies and labour unions, for example – are ill-equipped to deal with a new digital era of fast-paced monopolies. One of the real challenges for distributed trust is whether it can resist, or at least weather, market forces and human greed.

Ideas such as China's Social Credit System show how distributed networks of trust could become national networks of shame and interference, controlled by governments. And what has happened to those early utopian bitcoin miners? Mining power has ended up dangerously concentrated in China, at odds with the globalized ideals underlying bitcoin. The mass exchange of diverse ideas and the decentralization of information that we first envisioned the World Wide Web would bring us have happened, but so has a new kind of homophily and centralization – hyperlinks and hierarchies managing what we see and read – inside a handful of social networks.[12] It's as if the small local cafes where we talked and disagreed with strangers have been replaced by a chain of McCafes where we are given algorithmically determined food, regardless of what we might actually want. The consequence is that we have become vulnerable to digital concentrations of power. We want power handed back to the people, but what if it's handed to the wrong people? Or only some of the people? Or, worse still, only a few of the wrong people?

It would be easy to suggest that platforms should be owned and governed by their users but there is the issue of accountability. Even ideas, such as the DAO fund and bitcoin, that seek to challenge bureaucratic systems and powerful gatekeepers, seem to need top-down decision-makers at times. When the DAO fund went belly up, what were many people calling for? An individual: Ethereum founder Vitalik Buterin. Yes, he had to get a majority consensus from the user network to implement the hard fork solution; but it seems we still want to be able to throw up our hands and say 'It was his idea', or ask 'Who's in charge here?' For the moment, at least, we remain in a mindset that wants a benevolent leader, an ultimate decision-maker to take charge and fix the problem. The positive is that all these processes are far more transparent than ever before, as well as under mass observation and open to comment from everyone with a stake in them.

In truth, the creators and leaders of ideas built on distributed trust are not like traditional captains of the ship issuing orders in command-and-control model and manner. But as stuff happens – whether it's a mis-channelling of funds or an unpredictable mass murder associated with their platform – they are realizing they need to know about and take increased responsibility for everything and anything that could sink the ship. The unforeseen storms could be legal challenges, data breaches, safety questions, unethical behaviour, discrimination or even competitors. When the water starts rushing in, leaders have to act quickly and openly to plug the holes, and this can't always involve consensus building, although sometimes it can and should. Once the holes are plugged, they need to figure out how to fix the problem in the long-term in a 'We will do . . .' approach, engaging the wider community.

The third challenge of distributed trust is that many new technologies, from bots to blockchains, either anonymize people or attempt to remove entirely the need to trust another human. Yet it's humans, with all our wonderful kinks and mutations, who make trust possible. It's not technology or mathematics. When we trust an automated search engine over a human editor or when avatars or programmed algorithms serve as our managers, trust runs the risk of becoming static. What happens to human let-downs and surprises? These are how we learn to trust, and not trust. How else do we practise the skills of earning and rebuilding trust? Sometimes repairing trust requires a slow cure rate and the personal touch. It would be a shame to find ourselves in a world so automated that we depend solely on machines and algorithms to make decisions about whom to trust. That's a world apparently devoid of uncertainty, devoid of the colour and movement born of human imperfection, and, if we take our hands off the wheel too much, possibly even dangerous. As astronaut Dave Bowman famously found out when HAL 9000 went rogue in Kubrick's cult *2001: A Space Odyssey*, one of our key challenges is deciding where and when it is appropriate to make trust a matter of computer code.

When we look back in history, we can see that trust falls into distinct chapters. The first was *local*. The second was *institutional*. And the third, still very much in its infancy, is *distributed*. Like most inventions in their early stages, distributed trust will be messy, unpredictable and at times even dangerous. Researching and writing about the theory has sometimes felt remarkably similar to watching my two young children bounding around the house, pushing boundaries, constantly negotiating, feeling misunderstood, trying to figure out the rules they have to abide by and the ones they can ignore.

There is no simple answer to the question 'Who Can You Trust?' but we do know that ultimately it comes down to a human decision. Technology can help us make better and different choices, but in the end it's we who have to decide where to place our trust and who deserves it. It will require some care. Distributed trust needs us to allow space for a *trust pause*, an interval in which to stop and think before we automatically click, swipe, share and accept. To ask the right questions and to seek the right information that helps us to decide: is this person, information or thing worthy of my trust? What is it I'm trusting them to do or deliver? Each time we engage in that process, we are in our own small way taking responsibility for the kind of world we want to live in. We are exercising the power available to us all now at the press of a key. We are helping to preserve society's most precious and fragile asset, trust.

Glossary of 'Trust' Terms

Distributed Trust: Trust that flows laterally between individuals, enabled by networks, platforms and systems.

Institutional Trust: Trust that flows upwards to leaders, experts and brands, and runs through institutions and intermediaries such as courts, regulatory bodies and corporations (e.g. trusting your bank to safeguard your savings).

Local Trust: Trust that exists between members of small, local communities and rests *in* someone specific, someone we are familiar with.

Reputation: The overall opinion others have formed of you, based on past experiences and built up over time.

Reputation Capital: The value of your reputation across communities, networks and marketplaces; a measurement of how much an individual or community trusts you.

Reputation Trails: Data we leave behind about how we behave, or misbehave.

Trust: A confident relationship with the unknown.

Trust Blocker: Obstacles or deal breakers for people when it comes to trusting a new idea or each other (e.g. not believing self-driving cars will make the right safety decisions).

Trust Breach: An incident or act that causes an erosion of confidence in a person, product, system or company. Trust breaches may vary in severity.

Trust Deficit: A lack of trust in a business, institution or within a society that prevents it from functioning well.

Trust Engineers: People designing digital systems and networks that connect people and build or manipulate distributed trust.

Trust Gap: The void between the known and unknown.

Trust Influencers: People who can disproportionately influence a significant change in the way we do something or view something, and thus set new social norms.

Trust Leap: A trust leap occurs when we take a risk and do something new or in a fundamentally different way.

Trust Pause: An interval in which to stop and think before we automatically swipe, click and give our trust to someone.

Trust Scar: Created by a trust-busting incident, a scar against an institution, individual or brand that may take decades or generations to heal.

Trust Score: A system where all of an individual's behaviours are rated as either positive or negative and distilled into a single number as an indicator of their overall trustworthiness.

Trust Shift: The historical evolution of trust from local to institutional, and institutional to distributed.

Trust Signals: Clues or symbols that we knowingly or unknowingly use to decide whether or not another person is trustworthy.

Trust Stack: The three-step process of trusting the idea; then the platform; and finally, the other person (or in some instances a machine or robot).

Trust Vacuum: Created by a lack of trust in traditional experts, leaders and elites; an absence of trust that can create opportunities for malicious disruptors to occupy the space.

Trustworthy: Someone who is competent, reliable, benevolent and has integrity, and thus is worthy of our trust.

Acknowledgements

It's funny, I think writing books is a bit like having children; as soon as you have the first copy in your hands, you forget the immensely tough marathon in getting there. It takes a village to produce a book and I am deeply grateful for the advice, encouragement and support given by friends, family and colleagues along the way.

As for the book itself, I have three talented individuals to thank in particular: Mia de Villa, Phoebe Adler-Ryan and Fenella Souter.

Mia played an invaluable role as a research assistant, tracking down obscure papers and checking and rechecking hundreds of facts and transcripts. She was willing to do whatever it took to produce this book. Thank you, for everything.

A special thank you to Fenella who was instrumental in helping to make the stories in this book zing. She was an incisive critic throughout the process and her keen-eyed comments helped to make every page infinitely better.

I am immensely grateful to Phoebe, my assistant, for handling so much so well, and with diligent professionalism, care and grace. Thank you for being a joy to work with and an invaluable member of the team.

My heartfelt thanks to my brilliant literary agent, Toby Mundy, for continually playing the roles of thoughtful advisor, supportive coach and chief cheerleader. I am indebted to his fierce faith in the project (even in its unwieldy embryonic state) and for making this book happen. And thanks to David Roach for introducing me to Toby.

I'd like to extend a special thanks to my terrific editors at Penguin Portfolio, Fred Baty, Daniel Crewe and Martina O' Sullivan, who pushed the manuscript along in the most encouraging way possible. And the talented and world-class team behind the scenes: Nicola Evans, Mathias Lord, Lydia Yadi, Ellie Smith and John Stables. For her careful copyedit, thank you to Karen Whitlock. I was very fortunate to have David Over leading the marketing efforts around this book. Thank you for passionately embracing the ideas from the outset. Thanks also to Alex Elam and Sarah Scarlett and the rest of the dedicated international rights team at Penguin Portfolio.

I am also grateful to the first-rate team at Hachette, Public Affairs – led by Clive Priddle, Lindsay Fradkoff, and Jaime Leifer – who enthusiastically brought this book to readers in the United States.

My gratitude goes to the talented designers at Team Design: Amy Globus, John Clark and Devin Seger for the stunning internal illustrations. I am constantly amazed how you turn my chicken scratches into something beautiful. Making complex things simple is complex and you do it so well.

Thanks also to Caroline Baum, an inspirational friend and talented writer, who read through an early draft of the manuscript (it's because I trust you!).

A special thank you to Danny Stern, for doing everything a good speaking agent should do – only infinitely better. I am also grateful to the rest of the team at Stern Strategy Group for their committed enthusiasm for my ideas: Katie Balogh, Tara Baumgarten, Mel Blake, Stephanie Heckman, Whitney Jennings, Joseph Navatto, Susan Stern and Ania Trzepizur. And a big thank you to Nanette Moulton, Trish Stafford and Carol Pedersen and the rest of the team at Saxtons for supporting my engagements from the very first time I stepped on a stage.

Enormous thanks to all the talented people who graciously shared their time and wisdom with me, including: Judd Antin, Jessi Baker, Andrea Barrett, Savi Baveja, Joshua Browder, Leah Busque, Verena Butt d'Espous, Pierrick Caen, Juan Cartagena, Emily Castor, Stephen Cave, Coye Cheshire, Sean Conway, Ines Cormier, Rogier Creemers, Courtney Cregan, Nilesh Dalvi, Damien Detcherry,

Matt Faustman, Juliette Garside, Joe Gebbia, Logan Green, Alok Gupta, Elliot Hedman, José Ignacio Fernández, Angeli Jain, Husayn Kassai, Leanne Kemp, Federico Lalatta, David Lang, Brian Lathrop, James Martin, Frédéric Mazzella, Mark Stephen Meadows, Paolo Parigi, Lynn Perkins, Gerard Ryle, Anish Das Sarma, Ariel Schultz, Shivani Siroya, Ryann Wahl and Seth Weiner. (And to all the story contributors who asked to be kept anonymous.) I am immensely grateful for their trust in letting me share their experiences and stories.

James Coleman, Francis Fukuyama, Onora O'Neill and Robert Putnam, with their remarkable writings on trust, have been a touchstone of intellectual inspiration.

The faculty at Oxford University Saïd Business School have provided me with the opportunity to teach the unbelievable student community. A special thank you to Colin Mayer, Ian Rogan, Rupert Younger and Marc Ventresca for all their support. I also want to thank my students for their engagement, comments and for pushing me to find clearer answers. Pamela Hartigan, my former supervisor, to whom this book is dedicated, showed me, in innumerable ways, what it means to be a teacher. I benefited enormously from her wisdom and was inspired by her relentless faith in people.

Thanks to the colleagues and organizations who have frequently provided public platforms to test and improve ideas: Helen Goulden and the team at Nesta; Mairi Ryan and Matthew Taylor of the RSA; and Alain de Botton and the team at School of Life; Chris Stanley at *WOBI*; Jo Gray, Theo Chapman and the editors at the *AFR*; and David Rowan and Greg Williams at *WIRED*. A big thank you to Chris Anderson, Remo Giuffre, Bruno Giussani and Helen Walters at TED for letting me share my ideas from the ominous 'red circle'.

I am also thankful to Rohan Lund, Kyle Loades and the other members of the board at the NRMA for tolerating my schedule and absences. I am proud to be a part of such a trusted organization.

I am grateful to Kirsty de Garis for her research and input in the early stages of this book. Thank you for always being there. Many other friends and family, including Dana Ardi, Craig Baker, Tony

Botsman and Jonathan Simmons, who have listened (or I should say, at times, endured) obsessive conversations about ideas lingering in my mind.

A special thank you to Isabel 'one' and Isabel 'deux' for all their love and help with my kids. To all the teachers at Emanuel School, thanks for creating a wonderful and caring place.

My dear parents, Ruth and David, trusted me to share many personal family stories in this book. (I forgive you for leaving me in the care of a drug dealer!) An enormous thank you for all your boundless love, generosity and wisdom over the years. You are both remarkable people.

To my mother-in-law Wendy: my heartfelt thanks for all your love and support. You are the best nonna to the kids.

And of course, my two beautiful children, Jack and Grace, for being an infinite source of joy, pride and humility. Although they are both still young, they have taught me so much about trust and how it's a family's most precious asset.

Finally, my heartfelt thanks go to the person who knows me best, my husband, Chris. I am sure there were many times he did not want to talk about bots, blockchain and Breitbart, but he always listened. His endless patience, love and support made this book possible.

Rachel Botsman, June 2018

Notes

Introduction

1 'The Financial Crisis Inquiry Report', Financial Crisis Inquiry Commission United States of America, https://www.gpo.gov/fdsys/pkg/GPO-FCIC/pdf/GPO-FCIC.pdf, 25 February 2011, accessed 2 March 2017.

2 'Digital Wildfires in a Hyperconnected World', WEF Report 2016, http://www3.weforum.org/docs/Media/TheGlobalRisksReport2016.pdf, accessed 16 May 2017.

3 '2017 Edelman Trust Barometer', Edelman, http://www.edelman.com/trust2017/, accessed 2 March 2017.

4 '2017 Edelman Trust Barometer Reveals Global Implosion of Trust', Edelman, http://www.edelman.com/news/2017-edelman-trust-barometer-reveals-global-implosion/, accessed 15 January 2017.

5 Turing Pharmaceuticals' CEO, Martin Shrekli, was arrested for securities fraud on 18 December 2015 and he resigned the following day. As of June 2017, the case is still in trial. The CEO has denied fraud charges.

6 'The Billion Dollar Startup Club', *Wall Street Journal*, http://graphics.wsj.com/billion-dollar-club/, accessed 16 May 2017.

7 For Uber valuation see *Recode*, https://www.recode.net/2018/2/9/16996834/uber-latest-valuation-72-billion-waymo-lawsuit-settlement.

8 'Tinder reveals the most attractive jobs in the UK that make people swipe right', *Telegraph*, http://www.telegraph.co.uk/technology/2016/09/07/tinder-reveals-the-most-attractive-jobs-in-the-uk-that-make-peop/, 7 September 2016, accessed 30 May 2017.

1. *Trust Leaps*

1 See 'Alibaba IPO: Market values e-commerce giant at $231bn in enthusiastic opening day', *Guardian*, https://www.theguardian.com/business/live/2014/sep/19/alibaba-ipo-nyse-stock-price-live-updates, 12 November 2016, accessed 30 January 2017.
2 See 'Alibaba Lists on the NYSE', NYSE, https://www.nyse.com/network/article/Alibaba-Lists-on-the-NYSE, 19 September 2014, accessed 30 May 2017.
3 'Who rang Alibaba's IPO opening bell?', Offbeat China, http://offbeatchina.com/who-rang-alibabas-ipo-opening-bell, 22 September 2014, accessed 30 January 2017.
4 See '*Guanxi* vs networking: Distinctive configurations of affect- and cognition-based trust in the networks of Chinese vs American managers', Roy Chua, Michael Morris and Paul Ingram, *Journal of International Business Studies*, https://link.springer.com/article/10.1057/palgrave.jibs.8400422, 17 July 2008, accessed 30 November 2016.
5 'Entrepreneurial Masterclass: Alibaba founder Jack Ma interviewed by Charlie Rose', BizNews, http://www.biznews.com/interviews/2015/02/09/the-incredible-story-behind-alibabas-jack-ma-an-inspiration-that-will-span-generations/, 9 February 2015, accessed 4 May 2017.
6 'Alibaba, JD.com Locked in War of Words', *Forbes*, https://www.forbes.com/sites/ywang/2015/01/09/alibaba-jd-com-locked-in-war-of-words/#5cc3b3926658, 9 January 2015, accessed 30 May 2017.
7 'How Jack Ma Went From Being a Poor School Teacher to Turning Alibaba into a $US160 Billion Behemoth', Business Insider, http://www.businessinsider.com.au/the-story-of-jack-ma-founder-of-alibaba-2014-9?r=US&IR=T, 15 September 2014, accessed 30 May 2017.
8 'Entrepreneurial Masterclass: Alibaba founder Jack Ma interviewed by Charlie Rose', BizNews, http://www.biznews.com/interviews/2015/02/09/the-incredible-story-behind-alibabas-jack-ma-an-inspiration-that-will-span-generations/, 9 February 2015, accessed 4 May 2017.
9 'Meet Jack Ma, the man behind Alibaba', Australian Financial Review, http://www.afr.com/technology/meet-jack-ma-the-man-behind-alibaba-20140908-jeqeh, 10 September 2014, accessed 30 January 2017.

10 'Thirteen Fascinating Facts About the Man Behind the Largest IPO in History', NextShark, http://nextshark.com/13-fascinating-facts-about-the-man-behind-the-largest-ipo-in-history/, 23 September 2014, accessed 30 January 2017.
11 'Alibaba Group', Julie Wulf, Harvard Business School Case 710-436, http://www.hbs.edu/faculty/Pages/item.aspx?num=38507, 26 April 2010, last accessed 30 May 2017.
12 *Alibaba's World*, Porter Erisman, Macmillan (2015), p. 12.
13 'Jack Ma: U.S. Small Business Is Key to Alibaba's Growth', Bloomberg, https://www.bloomberg.com/news/articles/2015-06-09/u-s-small-business-key-to-alibaba-growth-outside-china-ma-says, 9 June 2015, accessed 30 January 2017
14 'Alibaba Group Announces March Quarter 2017 and Full Fiscal Year 2017 Results', Alibaba Group, http://www.alibabagroup.com/en/news/press_pdf/p170518.pdf, 18 May 2017, last accessed 28 February 2018.
15 'Squawk on the Street', CNBC, http://www.cnbc.com/2014/11/11/cnbc-exclusive-cnbc-transcript-alibaba-founder-executive-chairman-jack-ma-sits-down-with-cnbcs-david-faber-today-on-squawk-on-the-street.html, 11 November 2014, accessed 4 May 2017.
16 See 'Trust and Consequences', Eric Uslaner, University of Maryland, http://gvptsites.umd.edu/uslaner/commun.pdf, accessed 30 November 2016, p. 1.
17 See 'Gifts and Exchanges', essay by Kenneth Arrow, Princeton University Press (1982).
18 For the distinction between generalized and personal trust, see 'Trust and Finance', Paola Sapienza and Luigi Zingales, NBER Reporter Online, National Bureau of Economic Research, http://www.nber.org/reporter/2011number2/paola&luigi.html, June 2011, accessed 1 December 2016.
19 See *Trust and Power*, Niklas Luhmann, with an introduction by Gianfranco Poggi, Wiley (1979).
20 See *The Resolution of Conflict*, Morton Deutsch, Yale University Press (1973).
21 'Entrepreneurial Masterclass: Alibaba Founder Jack Ma Interviewed by Charlie Rose', BizNews, http://www.biznews.com/interviews/

2015/02/09/the-incredible-story-behind-alibabas-jack-ma-an-inspiration-that-will-span-generations/, 9 February 2015, accessed 30 May 2017.

22 'Social Networks, e-Commerce Platforms and the Growth of Digital Payment Ecosystems in China: What it Means for Other Countries', Better than Cash Alliance, https://www.betterthancash.org/tools-research/case-studies/social-networks-ecommerce-platforms-and-the-growth-of-digital-payment-ecosystems-in-china, 19 April 2017, accessed 28 February 2018.

23 See 'Evans Chamberlain Asset Management reports Ant Financial's $100 billion valuation', Evans Chamberlain Asset Management, https://www.prnewswire.com/news-releases/evans-chamberlain-asset-management-reports-ant-financials-100-billion-valuation--674169213.html, 15 February 2018, accessed 28 February 2018.

24 'Social Networks, e-Commerce Platforms and the Growth of Digital Payment Ecosystems in China: What it Means for Other Countries', Better than Cash Alliance, https://www.betterthancash.org/tools-research/case-studies/social-networks-ecommerce-platforms-and-the-growth-of-digital-payment-ecosystems-in-china, 19 April 2017, accessed 28 February 2018.

25 See 'Eight Ringers of Alibaba's IPO Opening Bell', Women of China, http://www.womenofchina.cn/womenofchina/html1/people/others/1409/1036-1.htm, 23 September 2014, accessed 30 January 2017.

26 *Alibaba's World*, Porter Erisman, Macmillan (2015), p. 44.

27 See 'Why Alibaba's CEO had to go', *Fortune*, http://fortune.com/2011/02/22/why-alibabas-ceo-had-to-go/, 22 February 2011, accessed 30 January 2017.

28 See 'Alibaba.com chief executive resigns', *Guardian*, https://www.theguardian.com/business/2011/feb/21/alibaba-chief-resigns-over-frauds, 21 February 2011, accessed 1 December 2016.

29 See 'The Market for Lemons: Quality Uncertainty and the Market Mechanism', George Akerlof, *Quarterly Journal of Economics*, 84, 3 (1970).

30 See 'The Maghribi traders: a reappraisal?' Avner Greif, *The Economic History Review*, https://web.stanford.edu/~avner/Greif_Papers/2012_Greif_long_ssrn_Maghribi.pdf, May 2012, accessed 21 February 2017.

31 'Reputation and Coalitions in Medieval Trade: Evidence on the Maghribi Traders', Avner Greif, *The Journal of Economic History* 49, 4 (1989), 857–83, https://web.stanford.edu/~avner/Greif_Papers/1989%20Greif%20JEH%201989.pdf, accessed 1 December 2016.

32 *Bowling Alone*, Robert Putnam, Touchstone Books by Simon and Schuster (2001).

33 See *Trust*, Francis Fukuyama, Free Press Paperbacks (1995), p. 10.

2. Losing Faith

1 'Syphilis Victims in US Study went untreated for 40 years', Jean Heller, *New York Times*, http://www.nytimes.com/1972/07/26/archives/syphilis-victims-in-us-study-went-untreated-for-40-years-syphilis.html, 26 July 1972, accessed 30 May 2017.

2 'The Tuskegee Experiment kept killing black people after it ended', *New York Magazine*, http://nymag.com/scienceofus/2016/06/tuskegee-experiment-mistrust.html, 15 June 2016, accessed 30 May 2017.

3 *Bad Blood: The Tuskegee Syphilis Experiment*, James H. Jones (Free Press, 1993).

4 See Jean Heller's article, 'Syphilis victims in US study went untreated for 40 years', *New York Times*, http://www.nytimes.com/1972/07/26/archives/syphilis-victims-in-us-study-went-untreated-for-40-years-syphilis.html, 26 July 1972, accessed 17 May 2017.

5 See *Statistical Monitoring of Clinical Trials: Fundamentals for Investigators*, Lemuel Moyé, Springer (2006), p. 6.

6 'Clinton's Apology', Tuskegee Study, http://www.cdc.gov/tuskegee/clintonp.htm, 16 May 1997, accessed 18 January 2017.

7 'Deaths: Final Data for 2010', S. L. Murphy, J. Xu and K. D. Kochanek, *National Vital Statistics Reports*, 61, 4 (2013).

8 The researchers used a statistical method called the 'triple-difference model' which crunched data from the General Social Survey, the National Interview Survey and the Centers for Disease and Control. See 'Tuskegee and the Health of Black Men', Marcella Alsan and Mari-

anne Wanamaker, National Bureau of Economic Research, http://www.nber.org/papers/w22323, June 2016, accessed 16 January 2017.

9 For two great references on the study and paper, see 'A Generation of Bad Blood', *The Atlantic*, http://www.theatlantic.com/politics/archive/2016/06/tuskegee-study-medical-distrust-research/487439, 17 June 2016, and 'Did Infamous Tuskegee Study Cause Lasting Mistrust of Doctors Among Blacks?', *New York Times*, http://www.nytimes.com/2016/06/18/upshot/long-term-mistrust-from-tuskegee-experiment-a-study-seems-to-overstate-the-case.html?_r=0, 17 June 2016, both accessed 30 May 2017.

10 See Joseph Ravenell's TED talk, 'How barbershops can keep men healthy', https://www.ted.com/talks/joseph_ravenell_how_barbershops_can_keep_men_healthy/transcript?language=en, February 2016, accessed 30 May 2017.

11 The authors are clear to note that the findings shed light on the men who responded to government surveys they analysed from 1972 and the years immediately following but do not necessarily reflect the attitudes of younger African-American men today. See 'Tuskegee and the Health of Black Men', Marcella Alsan and Marianne Wanamaker, National Bureau of Economic Research, http://www.nber.org/papers/w22323, June 2016, accessed 16 January 2017.

12 The Panama Papers, International Consortium of Investigative Journalists, https://panamapapers.icij.org/, accessed 16 May 2017.

13 To read about how the group effort worked with the ICIJ and journalists from around the world, see 'About the Panama Papers', Frederik Obermaier, Bastian Obermayer, Vanessa Wormer and Wolfgang Jaschensky, *Süddeutsche Zeitung*, http://panamapapers.sueddeutsche.de/articles/56febff0a1bb8d3c3495adf4/, accessed 16 January 2017.

14 'While one journalist is looking at Indian data, it might lead them to Brazil' and subsequent quotes from Gerard Ryle, author interview, 14 July 2016.

15 See 'Panama papers: Iceland PM Sigmundur Gunnlaugsson steps down', *Sydney Morning Herald*, http://www.smh.com.au/world/panama-papers-iceland-prime-minister-sigmundur-gunnlaugsson-offers-his-resignation-20160405-gnza99.html, 6 April 2016, accessed 4 March 2017.

16 See *The Great Divide: Unequal Societies and What We Can Do About Them*, Joseph Stiglitz, W. W. Norton (2015), and 'The Great Divide', speech by Andy Haldane given at New City Agenda Annual Dinner, London, on 18 May 2016.

17 For a very good paper on defining institutions, see 'What Are Institutions?', Geoffrey Hodgson, *Journal of Economic Issues*, March 2006, p. 394, http://www.geoffrey-hodgson.info/user/bin/whatareinstitu tions.pdf, accessed 16 January 2017.

18 *The Great Degeneration*, Niall Ferguson, Penguin Books (2014), p. 12.

19 The survey question asked goes like this: 'How much of the time do you think you can trust government in Washington to do what is right – a great deal, a fair amount, not very much, none at all?' See 'Narrow Majority Trust Government to Handle Foreign Problems', Gallup, http://news.gallup.com/poll/219662/narrow-majority-trust-gov ernment-handle-foreign-problems.aspx, accessed 28 February 2018.

20 See 'Americans' Confidence in Institutions Edges Up', Gallup, http:// news.gallup.com/poll/212840/americans-confidence-institutions-edges. aspx, accessed 28 February 2017.

21 Ibid.

22 Ibid.

23 Ibid.

24 Ibid.

25 Ibid.

26 Ibid.

27 Ibid.

28 See 'No Front-Runner Among Prospective Republican Candidates, Hillary Clinton in Control of Democratic Primary, Harvard Youth Poll Finds', Harvard Kennedy School, http://iop.harvard.edu/no-front-runner-among-prospective-republican-candidates-hillary-clinton-control-democratic-primary, accessed 16 January 2017.

29 See 'Veracity Index 2017', https://www.ipsos.com/sites/default/files/ ct/news/documents/2017-11/trust-in-professions-veracity-index-2017-slides.pdf, November 2017.

30 See 'Wall Street in Crisis: A Perfect Storm Looming', Labaton Sucharow's US Financial Services Industry Survey, http://www.labaton.com/

en/about/press/Wall-Street-Professional-Survey-Reveals-Widespread-Misconduct.cfm, 16 July 2013, accessed 16 January 2017.

31 See 'Carney puts banker's pay in spotlight after misconduct shockwaves', BBC, http://www.bbc.com/news/business-30079451, 17 November 2014, accessed 4 March 2017.

32 For a short profile of Andrew Haldane, see 'The central banker not afraid to be blunt', *TIME*, http://time.com/70833/andy-haldane-2014-time-100/, 23 April 2014, accessed 30 May 2017.

33 *Twilight of the Elites: America after Meritocracy*, Christopher Hayes, Broadway Books (2013), p. 102.

34 See 'VW labor leaders said to balk at big severance for former CEO', Bloomberg, https://www.bloomberg.com/news/articles/2015-09-28/vw-labor-leaders-said-to-balk-at-big-severance-for-former-ceo, 29 September 2015, accessed 4 March 2017.

35 'The Report of the Iraq Inquiry: Executive Summary', Iraq Inquiry, http://www.iraqinquiry.org.uk/media/247921/the-report-of-the-iraq-inquiry_executive-summary.pdf, p. 48, 6 July 2016, accessed 19 July 2017.

36 See 'News Feed FYI: Helping Make Sure You Don't Miss Stories From Your Friends', Facebook Newsroom, http://newsroom.fb.com/news/2016/06/news-feed-fyi-helping-make-sure-you-dont-miss-stories-from-friends/, 29 June 2016, accessed 4 March 2017.

37 See 'Yahoo! Tops Twitter as a Traffic Referral Source for Digital Publishers', Parse.ly, http://blog.parse.ly.com/post/3476/yahoo-tops-twitter-traffic-referral-source-digital-publishers/, 26 April 2016, accessed 5 March 2017.

38 Bill Bishop describes this phenomenon in the offline world in *The Big Sort: Why the Clustering of Like-Minded America Is Tearing Us Apart*, Mariner Books (2009). Yochai Benkler describes the online phenomenon in his book *The Wealth of Networks*, Yale University Press (2006).

39 *A Short History of Truth*, Julian Baggini, Hachette UK (2018), p. 102.

40 For statistics on where people get their news from, see 'News Use Across Social Media Platforms 2016', Pew Research Center, http://www.journalism.org/2016/05/26/news-use-across-social-media-platforms-2016/, 26 May 2016, accessed 30 May 2017.

41 The phrase 'filter bubble' was coined by MoveOn and Upworthy activist Eli Pariser, in his book *The Filter Bubble: What the Internet Is Hiding from You*, Penguin Press (2011).

42 See 'Obama Farewell Speech Transcript', *Los Angeles Times*, http://www.latimes.com/politics/la-pol-obama-farewell-speech-transcript-20170110-story.html, 10 January 2017, accessed 15 May 2017.

43 See '2016 Edelman Trust Barometer – Global Results', Edelman, http://www.slideshare.net/EdelmanInsights/2016-edelman-trust-barometer-global-results, accessed 17 January 2017.

44 See 'Beyond the Grand Illusion', Edelman, http://www.edelman.com/p/6-a-m/beyond-grand-illusion/, accessed 18 January 2016.

45 *The Prince*, Niccolò Machiavelli, Florence (1505), chapter 18.

46 See *Twilight of the Elites: America after Meritocracy*, Christopher Hayes, Crown Publishing Group (2012), p. 63.

47 See '5 Reasons Why Trump Will Win', Michael Moore, http://michaelmoore.com/trumpwillwin/, accessed 17 January 2017.

48 See transcript of interview with Faisal Islam and Michael Gove from 3 June 2016, 'EU in or out?', Sky News, https://corporate.sky.com/media-centre/media-packs/2016/eu-in-or-out-faisal-islam-interview-with-michael-gove-30616-8pm, accessed 16 May 2017.

49 See Michael Gove LBC Interview: 'Pro-EU Experts Like the Nazis, Says Gove', LBC, http://www.lbc.co.uk/hot-topics/eu-referendum/gove-compares-pro-eu-experts-to-nazis-132633/, 22 June 2016, accessed 19 July 2017.

50 See 'Word of the Year 2016 is . . .', Oxford Dictionary, https://en.oxforddictionaries.com/word-of-the-year/word-of-the-year-2016, accessed 17 January 2017.

51 New doublespeak dictionary, see Alain de Botton, https://twitter.com/alaindebotton/status/798623471735447553, 15 November 2016, accessed 12 December 2016.

52 See 'EU Referendum: Leave supporters trust ordinary "common sense" more than academics and experts', *Telegraph*, http://www.telegraph.co.uk/news/2016/06/16/eu-referendum-leave-supporters-trust-ordinary-common-sense-than/, 22 June 2016, accessed 5 March 2017.

53 See 'TV: Diverse Ventures in News and Public Affairs', *New York Times*, http://www.nytimes.com/1972/05/25/archives/tv-diverse-ventures-in-news-and-public-affairs.html?_r=0, 25 May 1972, accessed 5 March 2017.

54 To measure blog stats, the number of blog posts versus the number of blogs are measured because some blogs remain dormant/abandoned. See 'Blog posts written today', http://www.worldometers.info/blogs/, accessed 17 January 2017.

55 For the latest count of subreddits, according to Reddit, see 'Happy 10th birthday to us! Celebrating the best of 10 years of Reddit', Redditblog, https://redditblog.com/2015/06/23/happy-10th-birthday-to-us-celebrating-the-best-of-10-years-of-reddit/, 23 June 2015, accessed 5 March.

56 For the latest metrics for the subreddit 'Animals Without Necks' and also a useful site to look at numbers for other subreddits, see Reddit Metrics, http://redditmetrics.com/r/AnimalsWithoutNecks, accessed 17 January 2017.

57 See 'Removing harassing subreddits', Reddit, https://np.reddit.com/r/announcements/comments/39bpam/removing_harassing_subreddits/, accessed 17 January 2017.

58 For metrics on Reddit popularity, see 'How popular is Reddit', Alexa, http://www.alexa.com/siteinfo/reddit.com, accessed 17 January 2017, and 'Happy 10th birthday to us! Celebrating the best of 10 years of Reddit', Reddit blog, https://redditblog.com/2015/06/23/happy-10th-birthday-to-us-celebrating-the-best-of-10-years-of-reddit/, 23 June 2015, accessed 5 March 2017.

59 See 'TIFU by editing some comments and creating unnecessary controversy', Reddit, https://www.reddit.com/r/announcements/comments/5frg1n/tifu_by_editing_some_comments_and_creating_an/, 30 November 2016, accessed 17 January 2017.

60 See 'Petition demands Reddit CEO resign for editing Trump supporters' comments', IBTimes, http://www.ibtimes.co.uk/petition-demands-reddit-ceo-resign-editing-trump-supporters-comments-1593459, 25 November 2016, accessed 5 March 2017.

61 See 'Steve Huffman should step down as CEO of Reddit', Change.org, https://www.change.org/p/reddit-steve-huffman-should-step-down-as-ceo-of-reddit, accessed 22 February 2017.

3. Strangely Familiar

1 See 'BlaBlaCar valued at', IBIS Worldwide, http://ibisworldwide.com/2017/news/blablacar-valued-at-1-2bn/, accessed 4 May 2017.
2 The figure for the average trip taken is from 'Something to chat about', *The Economist*, http://www.economist.com/news/business/21676816-16-billion-french-startup-revs-up-something-chat-about, 24 October 2015, accessed 28 April 2017.
3 See 'BlaBlaCar has turned ride-sharing into a multi-million-euro business', *WIRED*, http://www.wired.co.uk/article/blablacar, 14 April 2015, accessed 28 April 2017.
4 Ibid.
5 See 'BlaBlaCar: Designing for Trust Between Strangers', Next, http://nextconf.eu/2013/09/blablacar-designing-for-trust-between-strangers/, 16 September 2013, accessed 28 April 2017. The '700 million trips' is from 'BlaBlaCar has turned ride-sharing into a multi-million-euro business', *WIRED*, http://www.wired.co.uk/article/blablacar, 14 April 2015, accessed 4 May 2017.
6 The company was renamed BlaBlaCar in 2011, ibid.
7 Ibid.
8 'What is the sharing economy?', BlaBlaCar, https://www.blablacar.com/blog/reinventing-travel/sharing-economy, accessed 17 May 2017.
9 For Eurostar number of travellers per year, see 'Passenger Numbers Stable', Eurostar, http://www.eurostar.com/uk-en/about-eurostar/press-office/press-releases/2016/passenger-numbers-stable-new-e320, accessed 16 May 2017. For BA number of travellers per year, see 'BA Traffic Stats', IAG Report 2016, http://www.iagshares.com/phoenix.zhtml?c=240949&p=irol-traffic, accessed 16 May 2017.
10 Food historians are divided on who invented the California Roll. See 'Will the Real Inventor of the California Roll Please Stand Up?', Grub

Street, http://www.grubstreet.com/2012/10/inventor-claims-california-roll-sushi.html, 24 October 2012, accessed 31 May 2017.

11 See 'Sushi Industry Statistics', Statisticbrain, http://www.statisticbrain.com/sushi-industry-statistics/, accessed 9 September 2016.

12 Steve Jobs believed that computers should be simple enough that a novice could master them based on instinct alone. In 2012, Apple moved away from this design philosophy because some of the old references like the reel-to-reel tape deck look of its podcast app was lost on younger smartphone users. See 'What is skeuomorphism?', BBC, http://www.bbc.com/news/magazine-22840833, 13 June 2003, accessed 28 April 2017.

13 *Critique of Pure Reason*, Immanuel Kant (1781).

14 Judd Antin's interview on Dollars to Donuts podcast, http://www.portigal.com/podcast/, 19 January 2016, accessed 30 May 2017.

15 Chris Sacca's interview on *This American Life*, http://www.thisamericanlife.org/radio-archives/episode/533/transcript, 9 June 2014, accessed 4 March 2017.

16 Inspired by Joe Gebbia's February 2016 TED talk, when he did the experiment with the audience. See 'How you can design for trust', TED, https://www.ted.com/talks/joe_gebbia_how_airbnb_designs_for_trust?language=en, accessed 17 May 2017.

17 'Olympic levels of trust' and subsequent quotes from Judd Antin, author interview, 20 July 2016.

18 Ibid.

19 For the details of Edward Jenner's development of the smallpox vaccination see Stefan Riedel, Baylor University Medical Center Proceedings, January 2005, 18(1): 21–25, 'Edward Jenner and the history of smallpox vaccination', NCBI, https://www.ncbi.nlm.nih.gov/pmc/articles/PMC1200696/#B10, accessed 16 May 2017.

20 'The Cow-Pock – or – the Wonderful Effects of the New Inoculation!' by James Gillray, 1870, from the *Catalogue of Political and Personal Satires in the Department of Prints and Drawings in the British Museum*, volume 11, London, http://www.britishmuseum.org/research/collection_online/collection_object_details.aspx?objectId=1638225&partId=1&people=18459&peoA=18459-1-7&page=1, last accessed 16 May 2017.

21 WHO, http://www.who.int/csr/disease/smallpox/en/, accessed 16 May 2017.

22 See 'Lancet retracts 12-year-old article linking autism to MMR vaccines', *CMAJ*, https://www.ncbi.nlm.nih.gov/pmc/articles/PMC2831678/, 4 February 2010, accessed 30 May 2017.

23 See 'Why Don't Parents Trust Vaccines?', Sharon Kaufman, Berkeley Wellness, http://www.berkeleywellness.com/healthy-community/contagious-disease/article/why-dont-parents-trust-vaccines, 6 April 2015, accessed 30 May 2017.

24 James Samuel Coleman believed that people act purposively towards their desired goals, usually acting to maximize utility with their goals and utilities shaped by values and preferences. See *Foundations of Social Theory*, James Samuel Coleman, Harvard University Press (1998).

25 See *Crossing the Chasm: Marketing and Selling High-Tech Products to Mainstream Customers*, Geoffrey Moore, HarperCollins (2006).

26 'You won't need a driver's license by 2040', *WIRED*, https://www.wired.com/2012/09/ieee-autonomous-2040/, 17 September 2009, accessed 16 May 2017.

27 'Autonomous vehicles face exactly the same challenge' and subsequent quotes from Brian Lathrop, author interview, 17 August 2016.

28 See 'Three-quarters of Americans "afraid" to ride in a self-driving vehicle', AAA Newsroom, http://newsroom.aaa.com/2016/03/three-quarters-of-americans-afraid-to-ride-in-a-self-driving-vehicle/, 1 March 2016, accessed 16 May 2017.

29 'Grandma freaks out self-driving Tesla – you will laugh', YouTube, https://www.youtube.com/watch?v=3-5QSZbcs-8, 15 April 2016, accessed 16 May 2017.

30 See Sebastian Thrun's TED talk, 'Google's driverless car', https://www.ted.com/talks/sebastian_thrun_google_s_driverless_car/transcript?language=en, March 2011, accessed 4 March 2017.

31 See 'Global status report on road safety 2013', World Health Organization, 2013, http://www.who.int/violence_injury_prevention/road_safety_status/2013/en/, accessed 31 May 2017.

32 'Self-Driving Cars Could Save 300,000 Lives Per Decade in America', *The Atlantic*, https://www.theatlantic.com/technology/archive/2015/09/

self-driving-cars-could-save-300000-lives-per-decade-in-america/407956/, 29 September 2015, accessed 31 May 2017.

33 See 'Connected and autonomous vehicles – the UK economic opportunity', KPMG, https://www.kpmg.com/BR/en/Estudos_Analises/artigosepublicacoes/Documents/Industrias/Connected-Autonomous-Vehicles-Study.pdf, March 2015, accessed 31 May 2017.

34 'Self-Driving Cars Could Save 300,000 Lives Per Decade in America', *The Atlantic*, https://www.theatlantic.com/technology/archive/2015/09/self-driving-cars-could-save-300000-lives-per-decade-in-america/407956/, 29 September 2015, accessed 31 May 2017.

35 See 'Loss Aversion in Riskless Choice: A Reference-Dependent Model', Amos Tversky and Daniel Kahneman, President and Fellows of Harvard College and the Massachusetts Institute of Technology, 1991, http://www3.uah.es/econ/MicroDoct/Tversky_Kahneman_1991_Loss%20aversion.pdf, accessed 16 May 2017.

36 See 'How the media screwed up the fatal Tesla accident', *Vanity Fair*, http://www.vanityfair.com/news/2016/07/how-the-media-screwed-up-the-fatal-tesla-accident, 7 July 2016, accessed 31 May 2017.

37 See 'A Tragic Loss', Tesla blog, https://www.tesla.com/en_AU/blog/tragic-loss, 30 June 2016, accessed 31 May 2017.

38 See 'Skype's first employee: How Taavet Hinrikus left Skype and founded TransferWise', YHP, http://yhponline.com/2012/03/20/taavet-hinrikus-transferwise/, 20 March 2012, accessed 31 May 2017.

39 See 'TransferWise wants to take over the world', TechCrunch, https://techcrunch.com/2015/12/07/transferwise-wants-to-take-over-the-world/, 7 December 2015, accessed 31 May 2017.

40 See 'Migration and Remittance Factbook 2016', http://siteresources.worldbank.org/INTPROSPECTS/Resources/334934-1199807908806/4549025-1450455807487/Factbookpart1.pdf, accessed 8 September 2016.

41 See *Influence: Science and Practice*, Robert Cialdini, Allyn and Bacon (2001).

42 Cialdini talked about this in the following paper summarizing his work on social proof, 'Dr Robert Cialdini and 6 principles of persuasion', http://www.influenceatwork.com/wp-content/uploads/2012/02/E_Brand_principles.pdf, accessed 16 September 2016.

43 The street corner experiment is described in 'Note on the Drawing Power of Crowds of Different Size', Stanley Milgram, Leonard Bickman and Lawrence Berkowitz, *Journal of Personality and Social Psychology*, 1969.
44 See *The Wisdom of Crowds*, James Surowiecki, Doubleday (2004), p. 43.
45 See TransferWise, https://transferwise.com/au, accessed 10 July 2018.

4. *Where Does the Buck Stop?*

1 This *GQ* article covering the shooting was used as reference in the details of the story, 'The Uber Killer: The Real Story of One Night of Terror', *GQ*, http://www.gq.com/story/the-uber-killer, 22 August 2016.
2 See 'Kalamazoo Uber Driver had a 4.73 Rating Before Shooting Rampage', *TIME*, http://time.com/4233148/kalamazoo-uber-driver-rating-shooting-rampage/, 23 February 2016, accessed 4 March 2017.
3 See 'Uber driver Jason Dalton charged with six counts of murder over alleged Michigan shooting rampage', ABC, http://www.abc.net.au/news/2016-02-23/uber-driver-charged-with-six-murders-in-us-shooting-rampage/7191728, 23 February 2016, accessed 4 March 2017.
4 See 'Passengers called 911 to report Uber driver before Kalamazoo shooting', Mashable, http://mashable.com/2016/02/22/kalamazoo-shooting-uber-driver-passengers/#osxstnSYHEqi, 23 February 2016, accessed 4 March 2017.
5 'Uber driver blamed, but no motive yet in killing spree', WZZM13, http://www.wzzm13.com/news/local/kalamazoo/uber-driver-blamed-but-no-motive-yet-in-killing-spree/50399251, 22 February 2016, accessed 4 March 2017.
6 See 'Kalamazoo Searches for Motive Spree That Killed 6', *New York Times*, http://www.nytimes.com/2016/02/22/us/kalamazoo-michigan-random-shootings.html?_r=0, 21 February 2016, accessed 4 March 2017.
7 See 'Uber failed to prioritize safety complaint on Kalamazoo suspect before shootings', *Guardian*, https://www.theguardian.com/us-news/2016/feb/22/kalamazoo-shooting-spree-jason-dalton-uber-ignored-safety-complaint, 23 February 2016, accessed 4 March 2017.

8 See 'The Truth About Uber's Background Checks', FastCompany, https://www.fastcompany.com/3050172/tech-forecast/the-truth-about-ubers-background-checks, 26 August 2015, accessed 4 March 2017.

9 See 'The Social Costs of Uber', *University of Chicago Law Review*, https://lawreview.uchicago.edu/page/social-costs-uber, 2015, accessed 15 May 2017.

10 See 'Reported list of incidents involving Uber and Lyft', Who's Driving You?, http://www.whosdrivingyou.org/rideshare-incidents, accessed 27 February 2017.

11 See footage and transcript of Travis Kalanick and Fawzi Kamel: 'A new video shows Uber CEO Travis Kalanick arguing with a driver over fares', Recode, https://www.recode.net/2017/2/28/14766964/video-uber-travis-kalanick-driver-argument, 28 February 2017, accessed 19 July 2017.

12 For Uber valuation see *Recode*, https://www.recode.net/2018/2/9/16996834/uber-latest-valuation-72-billion-waymo-lawsuit-settlement.

13 'More than 5 million' Uber passengers a day is taken from email correspondence from Alana Saltzman, Uber communications UK and Ireland, 6 June 2016.

14 See 'Uber takes "outrageous liberty" with drivers', *Sydney Morning Herald*, http://www.smh.com.au/business/the-economy/uber-takes-outrageous-liberty-with-drivers-20160413-go5m4f.html, 14 April 2016, accessed 30 May 2017.

15 See 'Uber lawsuits timeline: company ordered to pay out $161.9m since 2008', *Guardian*, https://www.theguardian.com/technology/2016/apr/13/uber-lawsuits-619-million-ride-hailing-app, 13 April 2016, accessed 16 May 2017.

16 See 'The battle is for the customer interface', TechCrunch, https://techcrunch.com/2015/03/03/in-the-age-of-disintermediation-the-battle-is-all-for-the-customer-interface/, 3 March 2015, accessed 4 March 2017.

17 See 'Horsemeat scandal: timeline', *Guardian*, https://www.theguardian.com/uk/2013/may/10/horsemeat-scandal-timeline-investigation, 11 May 2013, accessed 4 March 2017.

18 'Horsemeat scandal: where did the 29% horse in your Tesco burger come from', *Guardian*, https://www.theguardian.com/uk-news/2013/oct/22/horsemeat-scandal-guardian-investigation-public-secrecy, 22 October 2013, accessed 4 March 2017.
19 'Horsemeat scandal: Dutch trader found guilty and jailed', BBC, http://www.bbc.com/news/world-europe-32202995, 7 April 2015, accessed 19 July 2017.
20 See 'The Limits of Friendship', *New Yorker*, http://www.newyorker.com/science/maria-konnikova/social-media-affect-math-dunbar-number-friendships, 7 October 2014, accessed 4 March 2017.
21 See *The Evolution of Cooperation*, Robert Axelrod, Basic Books (1984).
22 Ibid.
23 See 'Bass Logo', Local History of Burton-on-Trent, http://www.burton-on-trent.org.uk/category/miscellany/bass-logo, accessed 27 February 2017.
24 'Get back in the box thought virus#6: brand as communication', Douglas Rushkoff, http://www.rushkoff.com/get-back-in-the-box-thought-virus-6-brand-as-communication/, accessed 27 February 2017.
25 'Global trust in advertising', Nielsen, http://www.nielsen.com/us/en/insights/reports/2015/global-trust-in-advertising-2015.html, 28 September 2015, accessed 31 May 2017.
26 See Joe Gebbia's TED talk, 'How Airbnb designs for trust', https://www.ted.com/talks/joe_gebbia_how_airbnb_designs_for_trust, February 2016.
27 'We have to create the conditions for a relationship to form between two people who have never met' and subsequent quotes from Joe Gebbia, author interview, 27 July 2016.
28 All further quotes from Alok Gupta, author interview 21 July 2016.
29 'Introducing Airbnb Verified', Airbnb blog, http://blog.airbnb.com/introducing-airbnb-verified-id/, 30 April 2013, accessed 27 February 2017.
30 'Digital Discrimination: The Case of Airbnb.com', B. Edelman and M. Luca, Harvard Business School, http://hbswk.hbs.edu/item/digital-discrimination-the-case-of-airbnb-com, 24 January 2014, accessed 31 May 2017.

31 'Racial Discrimination in the Sharing Economy: Evidence from a Field Experiment', B. Edelman, M. Luca and D. Svirsky, Harvard Business School, http://www.benedelman.org/publications/airbnb-guest-discrimination-2016-09-16.pdf, 16 September 2016, accessed 31 May 2017.

32 See 'Civil Rights Act of 1964 explained', http://civil.laws.com/civil-rights-act-of-1964, accessed 17 May 2017.

33 See 'Airbnb's Work to Fight Discrimination and Build Inclusion', Laura Murphy, Airbnb blog, http://blog.airbnb.com/wp-content/uploads/2016/09/REPORT_Airbnbs-Work-to-Fight-Discrimination-and-Build-Inclusion.pdf?3c10be, 8 September 2016, accessed 4 March 2017.

34 For @MiQL tweet see Twitter, https://twitter.com/MiQL/status/675834706529673216, last accessed 17 May 2017.

35 'Prejudices play out in the ratings we give – the myth of digital equality', *Guardian*, https://www.theguardian.com/technology/2017/feb/20/airbnb-uber-sharing-apps-digital-equality, 20 February 2017.

36 'Airbnb's Nondiscrimination Policy: Our Commitment to Inclusion and Respect', Airbnb, https://www.airbnb.com.au/help/article/1405/airbnb-s-nondiscrimination-policy-our-commitment-to-inclusion-and-respect, last accessed 1 March 2017.

37 'Online Trust, Trustworthiness, or Assurance?', MIT Press Journals, http://www.mitpressjournals.org/doi/abs/10.1162/DAED_a_00114?journalCode=daed#.WLKUqRJ96LI, 29 September 2011, accessed 31 May 2017.

38 Quotes from Coye Cheshire, author interview, 16 November 2016.

39 See 'Your car is a giant computer – and it can be hacked', CNN Money, http://money.cnn.com/2014/06/01/technology/security/car-hack/index.html, 2 June 2014, accessed 4 May 2017.

40 'Codebases: Millions of lines of code', Information is beautiful, http://www.informationisbeautiful.net/visualizations/million-lines-of-code/, accessed 4 May 2017.

41 'Social Clicks: What and Who Gets Read on Twitter?' Maksym Gabielkov, Arthi Ramachandran, Augustin Chaintreau, Arnaud Legout, *ACM SIGMETRICS / IFIP Performance 2016*, June 2016,

Antibes Juan-les-Pins, France, https://hal.inria.fr/hal-01281190, accessed 31 May 2017.

42 '6 in 10 of you will share this link without reading it, a new, depressing study says', *Washington Post*, https://www.washingtonpost.com/news/the-intersect/wp/2016/06/16/six-in-10-of-you-will-share-this-link-without-reading-it-according-to-a-new-and-depressing-study/?utm_term=.d9b38e787de3, 16 June 2016, accessed 4 March 2017.

43 Twitter, https://twitter.com/seanspicer?lang=en, accessed 16 May 2017.

44 See 'Transcript of Simon Sinek's Millennials in the Workplace Interview', Ochen, http://ochen.com/transcript-of-simon-sineks-millennials-in-the-workplace-interview, 4 January 2017.

45 'Facebook Tinkers with Users' Emotions in News Feed Experiment, Stirring Outcry', *New York Times*, https://www.nytimes.com/2014/06/30/technology/facebook-tinkers-with-users-emotions-in-news-feed-experiment-stirring-outcry.html?_r=0, 30 June 2014, accessed 4 March 2017.

46 'Experimental evidence of massive-scale emotional contagion through social networks', Cornell, https://cornell.app.box.com/v/fbcontagion, 25 March 2014, accessed 4 March 2017.

47 'Data Policy', Facebook, https://www.facebook.com/policy.php, accessed 27 February 2017.

48 'Facebook and Engineering the Public: It's not what's published (or not), but what's done', Medium, https://medium.com/message/engineering-the-public-289c91390225#.d1x3rghwy, 30 June 2014, accessed 30 May 2017.

49 'Here's what you need to know about that Facebook experiment that manipulated your emotions', Gigaom, https://gigaom.com/2014/06/30/heres-what-you-need-to-know-about-that-facebook-experiment-that-manipulated-your-emotions/, 30 June 2014, accessed 4 March 2017.

50 'The Trust Engineers', RadioLab, http://www.radiolab.org/story/trust-engineers/, 9 February 2015.

51 See *Weapons of Math Destruction*, Cathy O'Neil, Crown (2016), p. 183.

52 'The evolving role of news on Twitter and Facebook', Pew Research Center, http://www.journalism.org/2015/07/14/the-evolving-role-of-news-on-twitter-and-facebook/, 14 July 2015, accessed 4 March 2017.

53 'Hyperpartisan Facebook Pages are Publishing False and Misleading Information at an Alarming Rate', BuzzFeed News, https://www.buzzfeed.com/craigsilverman/partisan-fb-pages-analysis?utm_term=.qkPyQLqm8#.lcVrz58d2, 20 October 2016, accessed 1 March 2017.

54 'This Analysis Shows How Viral Fake Election News Stories Outperformed Real News on Facebook', BuzzFeed News, https://www.buzzfeed.com/craigsilverman/viral-fake-election-news-outperformed-real-news-on-facebook?utm_content=buffer20bf6&utm_medium=social&utm_source=twitter.com&utm_campaign=buffer&utm_term=.apz6rG5dQ#.habG6qdrR, 17 November 2016, accessed 17 November 2016.

55 'Eight Revealing Moments from the Second day of Russia Hearings', *WIRED*, https://www.wired.com/story/six-revealing-moments-from-the-second-day-of-russia-hearings/, 11 January 2017, last accessed 7 March 2018.

56 'Spinoff: Whatever the Reports about Russian Trolls Buying Ads is Initially, it's Way, Way Worse', *TechDirt*, https://www.techdirt.com/articles/20171031/11090138521/spinoff-whatever-reports-about-russian-trolls-buying-ads-is-initially-way-way-worse.shtml, 31 October 2017, last accessed 7 March 2018.

57 'Facebook estimates 126 million people were served content from Russia-linked pages', CNN, http://money.cnn.com/2017/10/30/media/russia-facebook-126-million-users/index.html, 31 October 2017, last accessed 7 March 2018.

58 'Mark Zuckerberg says fake news on Facebook could not have influenced the 2016 election', Quartz, https://qz.com/836079/mark-zuckerberg-says-fake-news-on-facebook-could-not-have-influenced-the-2016-election-of-donald-trump/, 13 November 2016, accessed 17 November 2016.

59 'Building Global Community', Mark Zuckerberg, Facebook, https://www.facebook.com/notes/markzuckerberg/buildingglobalcommunity/10103508221158471/?pnref=story, accessed 17 February 2017.

60 '2018 Edelman Trust Barometer Global Report', Edelman, January 2018, last accessed 7 March 2018.

61 Quotes from Joe Gebbia, author interview, 27 July 2016.

5. But She Looked the Part

1 See *Trust in Society*, ed. Karen S. Cook, Russell Sage Foundation (2003), 'Chapter 5: Trust in Signs', Michael Bacharach and Diego Gambetta.

2 I keep going back to Onora O'Neill's TED talk throughout the book. See Onora O'Neill, 'What we don't understand about trust', https://www.ted.com/talks/onora_o_neill_what_we_don_t_understand_about_trust/transcript?language=en, September 2013, accessed 31 May 2017.

3 'Trust, Trustworthiness and Transparency', European Foundation Centre, http://www.efc.be/human-rights-citizenship-democracy/trust-trustworthiness, 3 December 2015, accessed 30 May 2017.

4 *Trust and Trustworthiness*, Russell Hardin, Russell Sage Foundation (2002).

5 'Monster Mensch', *New York Magazine*, http://nymag.com/news/businessfinance/54703/, 22 February 2009, accessed 30 May 2017.

6 Onora O'Neill's TED talk, 'What we don't understand about trust', https://www.ted.com/talks/onora_o_neill_what_we_don_t_understand_about_trust/transcript?language=en, September 2013, accessed 31 May 2017.

7 *The Truth about Trust*, David DeSteno, Plume (2015), 'Chapter 6: Can I Trust You? Unlocking the Signals of Trustworthiness'.

8 UK-based executive communications consultancy The Aziz Corporation rated UK accents for their business appeal, and Scottish accents scored highly with 43 per cent of respondents saying that the Scottish speaker sounded successful; 40 per cent found the speaker to be hard-working and reliable and 31 per cent found them the most trustworthy. See 'Scots Accent Favored for Call Centers', WallStreet & Technology, http://www.wallstreetandtech.com/careers/scots-accent-favored-for-call-centers-/d/d-id/1256416, 26 March 2004, accessed 30 May 2017.

9 This is work from earlier scientists studying making judgements based on facial appearance. See 'How your looks betray your personality', Roger Highfield, Richard Wiseman and Rob Jenkins, *New Scientist*, https://www.newscientist.com/article/mg20126957-300-

how-your-looks-betray-your-personality/, 11 February 2009, accessed 30 May 2017.

10 See 'Static and Dynamic Facial Cues Differentially Affect the Consistency of Social Evaluation', *Personality and Social Psychology Bulletin*, http://psp.sagepub.com/content/early/2015/06/12/0146167215591495.abstract, 22 May 2015, accessed 30 May 2017.

11 See 'So you think you can be a hair braider', *New York Times*, http://www.nytimes.com/2012/06/17/magazine/so-you-think-you-can-be-a-hair-braider.html?_r=2&ref=magazine&pagewanted=all, 12 June 2012, accessed 30 May 2017.

12 Quote from Seth Weiner, author interview, 23 November 2016.

13 Trust game experiment conducted by Li Huang and Keith Murnighan. See 'A Trusted Name', Kellogg Insight, http://insight.kellogg.northwestern.edu/article/a_trusted_name, 1 June 2011, accessed 16 May 2017.

14 Ibid.

15 Jason Tanz wrote about how sharing-economy platforms like Lyft and Airbnb got Americans to interact and trust one another and not see meeting strangers as a huge risk. See 'How Airbnb and Lyft finally got Americans to trust each other', *WIRED*, https://www.wired.com/2014/04/trust-in-the-share-economy/, 23 April 2014, accessed 16 May 2017.

16 If you're logged on to Facebook and check this blog, you'll be able to see your degree of separation from everyone else. Last checked, Mark Zuckerberg has 3.17 degrees of separation; Sheryl Sandberg has 2.92. See 'Three and a half degrees of separation', Research at Facebook, https://research.facebook.com/blog/three-and-a-half-degrees-of-separation/, 4 February 2016, accessed 4 March 4 2017.

17 Sanjay Nazerali, chief strategy officer at Carat Global wrote a great post on the difference between faith and experience that helped me think through this point. See 'Faith vs experience: Building trust in the digital age', MMG, http://mandmglobal.com/faith-vs-experience-building-trust-in-the-digital-age/, 21 September 2015.

18 'A Question of Trust', Reith Lectures 2002, BBC Radio 4, http://www.bbc.co.uk/radio4/reith2002/, accessed 30 May 2017.

19 Traits of trustworthiness have been reviewed extensively in trust literature. Even though leading trust researchers propose different characteristics that are responsible for trust, there are three common characteristics that appear often: competence (sometimes labelled as ability or expertise), honesty (sometimes labelled as intention, integrity, benevolence, loyalty or goodwill) and reliability (sometimes labelled as dependability or availability). See 'An Integrative Model of Organizational Trust', The Academy of Management Review, Roger Mayer, James Davis and David Schoorman study, https://www.jstor.org/stable/258792?seq=1#page_scan_tab_contents, July 1995, accessed 30 May 2017.

20 See *Trust and Trustworthiness*, Russell Hardin, Russell Sage Foundation (2002), p. 6.

21 See 'Graduates are stretching the truth to get work in uncertain economic times', Pre-employment Screening blog, http://pre-employment-screening.blogspot.com.au/2009/08/graduates-are-stretching-truth-to-get.html, 7 August 2009, accessed 4 March 2017.

22 Quotes from Lynn Perkins, author interview, 25 August 2016.

23 Statistic provided by Lynn Perkins, author interview, 25 August 2016.

24 'People rate everything now' and subsequent quotes from Andrea Barrett, author interview, 25 August 2016.

25 See 'Web Archive: Cartoon Captures Spirit of the Internet', *New Yorker*, http://web.archive.org/web/20141030135629/http://www.nytimes.com/2000/12/14/technology/14DOGG.html, accessed 16 May 2017.

26 To see a copy of the cartoon go to 'The Joy of Tech', Me.Me, https://me.me/i/the-joy-of-tech-in-the-1990s-on-the-internet-11890719, accessed 16 May 2017.

27 This and subsequent quotes from Savi Baveja, author interview, 17 November 2016.

28 From Anish Das Sarma, author interview, 28 September 2016.

29 'Bad data fouls background checks', *WIRED*, https://www.wired.com/2005/03/bad-data-fouls-background-checks/, 11 March 2005, accessed 31 May 2017.

30 See 'Indefinite punishment and the criminal record: stigma reports among expungement-seekers in Illinois', http://onlinelibrary.wiley.com/wol1/doi/10.1111/1745-9125.12108/full, 8 June 2016, accessed 31 May 2017.

6. Reputation is Everything, Even in the Dark

1 How we view e-commerce now is so different from how it originally started, so there will be a few interpretations of how 'sold' is defined. A few contenders for the first thing sold on the internet are pizza, a Sting CD and weed. See *What the Dormouse Said: How the Sixties Counterculture Shaped the Personal Computer Industry*, John Markoff, Penguin Books (2005).

2 'New "Google" for the dark web makes buying dope and guns easy', *WIRED*, https://www.wired.com/2014/04/grams-search-engine-dark-web/, 17 April 2014, accessed 3 March 2017.

3 The FBI estimated that Silk Road had 13,000 drug listings and had processed approximately $1.2 billion. See *The Dark Net*, Jamie Bartlett, Windmill Books (2015), p. 136.

4 Ibid.

5 'Bitcoin "exit scam": deep-web market operators disappear with $12m', *Guardian*, https://www.theguardian.com/technology/2015/mar/18/bitcoin-deep-web-evolution-exit-scam-12-million-dollars, 19 March 2015, accessed 3 March 2017.

6 See 'The dark web's top drug market evolution, just vanished', *WIRED*, https://www.wired.com/2015/03/evolution-disappeared-bitcoin-scam-dark-web/, 18 March 2015, accessed 3 March 2017.

7 'Technology could be used to transform an illicit drugs trade' and subsequent quotes from James Martin, author interview, 22 September 2016.

8 See 'The Global Drug Survey 2014 Findings', Global Drug Survey, https://www.globaldrugsurvey.com/past-findings/the-global-drug-survey-2014-findings, accessed 4 May 2017.

9 See 'The Global Drug Survey 2016 Findings', Global Drug Survey, https://www.globaldrugsurvey.com/past-findings/the-global-drug-survey-2016-findings/, accessed 4 May 2017.

10 See 'Taking Stock of the Online Drugs Trade', Rand Corporation, http://www.rand.org/randeurope/research/projects/online-drugs-trade-trafficking.html, 27 October 2016, accessed 3 March 2017.

11 'Shedding Light on the Dark Web', *The Economist*, http://www.economist.com/news/international/21702176-drug-trade-moving-street-online-cryptomarkets-forced-compete, 16 July 2016, accessed 3 March 2017.
12 'The internet and drug markets', European Monitoring Centre for Drugs and Drug Addiction, http://www.emcdda.europa.eu/system/files/publications/2155/TDXD16001ENN_FINAL.pdf, February 2016, accessed 30 May 2017.
13 James Martin, author interview, 22 September 2016.
14 '"Fair Trade" Cocaine and "Conflict-free" Opium: The Future of Online Drug Marketing', The Conversation, http://theconversation.com/fair-trade-cocaine-and-conflict-free-opium-the-future-of-online-drug-marketing-30127, 12 August 2014, accessed 4 May 2017.
15 See *The Dark Net*, Jamie Bartlett, Windmill Books (2015), p. 162.
16 For statistics regarding positive reviews left on the darknet, see 'Traveling the Silk Road: A measurement analysis of a large anonymous online marketplace', Nicolas Christin, Carnegie Mellon University, https://www.cylab.cmu.edu/files/pdfs/tech_reports/CMUCyLab12018.pdf, 28 November 2012, accessed 31 May 2017.
17 See 'Trust among strangers in internet transactions', Paul Resnick and Richard Zeckhauser, http://cseweb.ucsd.edu/groups/csag/html/teaching/cse225s04/Reading%20List/E-bay-Empirical-BodegaBay.pdf, 20 September 2016, accessed 31 May 2017.
18 For changes in Airbnb review system, see 'Building for Trust', Medium, AirbnbEng, https://medium.com/airbnb-engineering/building-for-trust-503e9872bbbb#.s7872icvv, 15 March 2016, accessed 31 May 2017.
19 For Airbnb rating details, see 'How do star ratings work?', Airbnb, https://www.airbnb.com.au/help/article/1257/how-do-star-ratings-work?topic=207, accessed 28 October 2016.
20 'Evolution of the Future', Robert Axelrod, http://www.eleutera.org/wp-content/uploads/2015/07/The-Evolution-of-Cooperation.pdf, 1984.
21 I have spoken at length about what reputation is and how it is related to trust with Juan Cartegena, the founder of Traity. He was the first person to point out to me that reputation is best described as a 'risk premium' (not as an asset or currency) and that it has a price elasticity.

22 See 'Use of Silk Road, the online drug marketplace, in the United Kingdom, Australia and the United States', M. J. Barratt, J. A. Ferris, and A. R. Winstock, https://www.ncbi.nlm.nih.gov/pubmed/24372954, 2014, accessed 30 May 2017.

23 'Amazon targets 1,114 "fake reviewers" in Seattle lawsuit', BBC News, http://www.bbc.com/news/technology-34565631, 18 October 2015, accessed 31 May 2017.

24 'Historian Orlando Figes admits posting Amazon reviews that trashed rivals', *Guardian*, https://www.theguardian.com/books/2010/apr/23/historian-orlando-figes-amazon-reviews-rivals, 23 April 2010, accessed 3 March 2017.

25 'Orlando Figes: Historian admits to writing anonymous reviews on Amazon', *Telegraph*, http://www.telegraph.co.uk/culture/books/booknews/7622877/Orlando-Figes-Historian-admits-to-writing-anonymous-reviews-on-Amazon.html, 24 April 2010, accessed 19 June 2017.

26 'Poison pen reviews were mine, confesses historian Orlando Figes', *Guardian*, https://www.theguardian.com/books/2010/apr/23/poison-pen-reviews-historian-orlando-figes, 23 April 2010, accessed 19 July 2017.

27 'Some online reviews are too good to be true; Cornell computers spot "opinion spam"', http://www.news.cornell.edu/stories/2011/07/cornell-computers-spot-opinion-spam-online-reviews, 25 July 2011, accessed 31 May 2017.

7. Rated: Would Your Life Get a Good Trust Score?

1 'Planning Outline for the Construction of a Social Credit System', China Copyright and Media, translated by Rogier Creemers, https://chinacopyrightandmedia.wordpress.com/2014/06/14/planning-outline-for-the-construction-of-a-social-credit-system-2014-2020/, accessed 3 March 2017.

2 Ibid.

3 See 'China rates its own citizens – including online behaviour', *Volkskrant*, http://www.volkskrant.nl/buitenland/china-rates-its-own-

citizens-including-online-behaviour~a3979668/, 25 April 2015, accessed 4 May 2017.

4 'Big Brother Ranking for All: How Would Your Life Rate?', News.com.au, http://www.news.com.au/lifestyle/real-life/big-brother-ranking-for-all-how-would-your-life-rate/news-story/53928e0017a582e16acfa792bf51a496, 9 October 2015, accessed 3 March 2017.

5 See 'How China Wants to Rate Its Citizens', *New Yorker*, http://www.newyorker.com/news/daily-comment/how-china-wants-to-rate-its-citizens, 3 November 2015, accessed 30 April 2017.

6 'Use Big Data Thinking and Methods to Enhance the Government's Governing Capacity', China Copyright and Media, edited by Rogier Creemers, https://chinacopyrightandmedia.wordpress.com/2016/07/12/use-big-data-thinking-and-methods-to-enhance-the-governments-governing-capacity/, 12 July 2016, accessed 3 March 2017.

7 How do people get their Sesame Score? See 'Ant Financial Unveils China's First Credit-Scoring System Using Online Data', Alibaba Group, http://www.alibabagroup.com/en/news/article?news=p150128, 28 January 2015, accessed 15 March 2018.

8 See 'Ant Financial Subsidiary Starts Offering Individual Credit Scores', Caixin Online, https://www.caixinglobal.com/2015-03-02/101012655.html, accessed 15 March 2018.

9 Ibid.

10 *The Circle*, Dave Eggers, Penguin Books (2014), p. 367.

11 Ibid., p. 303.

12 See 'China wants to give all of its citizens a score – and their rating could affect every area of their lives', *Independent*, http://www.independent.co.uk/news/world/asia/china-surveillance-big-data-score-censorship-a7375221.html, 22 October 2016, accessed 30 April 2017.

13 See 'Mainland credit-rating network takes shape', *China Daily Asia*, http://www.chinadailyasia.com/business/2015-06/09/content_15274221.html, 9 June 2015, accessed 30 April 2017.

14 See 'In China, Your Credit Score is Now Affected By Your Political Opinions – And Your Friends' Political Opinions', Privacy News Online, https://www.privateinternetaccess.com/blog/2015/10/in-china-

your-credit-score-is-now-affected-by-your-political-opinions-and-your-friends-political-opinions/, 3 October 2015, accessed 4 March 2017.

15 Quotes from Rogier Creemers, author interview, 24 November 2016.

16 See 'In China, Your Credit Score is Now Affected By Your Political Opinions – And Your Friends' Political Opinions', Privacy News Online, https://www.privateinternetaccess.com/blog/2015/10/in-china-your-credit-score-is-now-affected-by-your-political-opinions-and-your-friends-political-opinions/, 3 October 2015, accessed 4 March 2017.

17 See 'China's "social credit": Beijing sets up huge system', BBC, http://www.bbc.com/news/world-asia-china-34592186, 26 October 2015, accessed 4 March 2017.

18 Cathy O'Neil covers this point in her excellent book about how algorithms are increasingly regulating people. See *Weapons of Math Destruction*, Cathy O'Neil, Crown (2016).

19 *Super Sad True Love Story*, Gary Shteyngart, Random House (2010).

20 'Will Social Media Make Us Anti-Social? A Talk With Gary Shteyngart', *The Atlantic*, http://www.theatlantic.com/business/archive/2011/10/will-social-media-make-us-anti-social-a-talk-with-gary-shteyngart/247373/, 26 October 2011, accessed 2 March 2017.

21 See 'China rates its own citizens – including online behaviour', *Volkskrant*, http://www.volkskrant.nl/buitenland/china-rates-its-own-citizens-including-online-behaviour~a3979668/, 25 April 2015, accessed 4 May 2017.

22 See 'China's "social credit": Beijing sets up huge system', BBC, http://www.bbc.com/news/world-asia-china-34592186, 26 October 2015, accessed 2 March 2017.

23 See 'Orwellian Dystopia or Trustworthy Nation? Get the Facts on China's Social Credit System', Advox Global Voices, https://advox.globalvoices.org/2016/01/08/orwellian-dystopia-or-trustworthy-nation-get-the-facts-on-chinas-social-credit-system/, 8 January 2016, accessed 3 March 2017.

24 See 'From the end of sesame credit "not the same" data source', 21jingji, http://m.21jingji.com/article/20150617/0c3b29fd50dd0a4a4b2f9d9e94f9cb99.html, 17 June 2015, accessed 30 May 2017.

25 'State Council Guiding Opinions concerning Establishing and Perfecting Incentives for Promise-keeping and Joint Punishment Systems for Trust-Breaking, and Accelerating the Construction of Social Sincerity', China Copyright and Media, edited by Rogier Creemers, https://chinacopyrightandmedia.wordpress.com/2016/05/30/state-council-guiding-opinions-concerning-establishing-and-perfecting-incentives-for-promise-keeping-and-joint-punishment-systems-for-trust-breaking-and-accelerating-the-construction-of-social-sincer/, 18 October 2016, accessed 8 March 2017.

26 Ibid. See also 'China's New Tool for Social Control: A Credit Rating for Everything', *Wall Street Journal*, https://www.wsj.com/articles/chinas-new-tool-for-social-control-a-credit-rating-for-everything-1480351590, 28 November 2016, accessed 31 May 2017.

27 See 'Charlie Brooker: the dark side of our gadget addiction', *Guardian*, https://www.theguardian.com/technology/2011/dec/01/charlie-brooker-dark-side-gadget-addiction-black-mirror, 2 December 2011, accessed 30 April 2017.

28 See 'Lessons from Luciano Floridi, the Google Philosopher', Radio National, http://www.abc.net.au/radionational/programs/philosopherszone/lessons-from-luciano-floridi-the-google-philosopher/6497872, 26 May 2015, accessed 4 March 2017.

29 See *The Fourth Revolution*, Luciano Floridi, Oxford University Press (2014).

30 Bret Easton Ellis's article in the *New York Times* about how everyone today is setting themselves up to be branded, targeted and data-mined online shaped some parts of this chapter. See 'Bret Easton Ellis on Living in the Cult of Likability', *New York Times*, http://www.nytimes.com/2015/12/08/opinion/bret-easton-ellis-on-living-in-the-cult-of-likability.html?_r=0, 8 December 2015.

31 For the proceedings of the 2010 ACM Conference on Computer Supported Cooperative Work see 'Is it really about me? Message content in social awareness streams', Association for Computing Machinery, 2010, pp. 189–92.

32 'Disclosing information about the self is intrinsically rewarding', *PNAS*, http://www.pnas.org/content/109/21/8038.full, 22 May 2012, accessed 30 May 2017.

33 Peeple, a people-rating app dubbed the 'Yelp for humans' was first publicized in October 2015 but faced heavy criticism. It relaunched in March 2016 after a few changes to the app. See 'Remember Peeple? It's back, and launching on Monday', *WIRED*, http://www.wired.co.uk/article/peeple-social-reputation-app-launched-released-download, 4 March 2016, accessed 16 May 2017.

34 'Everyone you know will be able to rate you on the terrifying "Yelp for people" – whether you want them to or not', *Washington Post*, https://www.washingtonpost.com/news/the-intersect/wp/2015/09/30/everyone-you-know-will-be-able-to-rate-you-on-the-terrifying-yelp-for-people-whether-you-want-them-to-or-not/, 30 September 2015, accessed 9 November 2016.

35 See 'Peeple Watching Webisode 1 – Building the people app in SF', YouTube, https://www.youtube.com/watch?v=6YrLEL6U504, 5 October 2015, accessed 16 May 2017.

36 'China penalizes 6.7m debtors with travel ban', *Financial Times*, https://www.ft.com/content/ceb2a7f0-f350-11e6-8758-6876151821a6, accessed 15 February 2017.

37 See '4.9 mln people with poor credit record barred from taking planes', Ecns.cn, http://www.ecns.cn/2016/11-03/232618.shtml, 11 March 2016, accessed 15 February 2017.

38 See 'Around the House: 50 years ago, an idea and $800 led to today's FICO scores', *San Gabriel Valley Tribune*, http://www.sgvtribune.com/business/20160624/around-the-house-50-years-ago-an-idea-and-800-led-to-todays-fico-scores, 24 June 2016, accessed 15 November 2017.

39 Refer to 'Fair and Accurate Credit Transactions Act of 2003', US Government Publishing Office, https://www.gpo.gov/fdsys/pkg/PLAW-108publ159/html/PLAW-108publ159.htm, accessed 10 May 2017.

40 See 'Credit score statistics', NASDAQ, http://www.nasdaq.com/article/credit-score-statistics-cm435901, 23 January 2015, accessed 4 May 2017.

41 See 'China's "social credit": Beijing sets up huge system', BBC, http://www.bbc.com/news/world-asia-china-34592186, 26 October 2015, accessed 31 May 2017.

42 For annual economic loss caused by lack of credit information, see 'China's social credit system gains momentum', ChinaDaily.com, http://www.chinadaily.com.cn/china/2014-08/22/content_18472094.htm, 22 July 2014, accessed 4 August 2017.

43 'Global trade in fake goods worth nearly half a trillion dollars a year – OECD & EUIPO', OECD, http://www.oecd.org/industry/global-trade-in-fake-goods-worth-nearly-half-a-trillion-dollars-a-year.htm, 18 April 2016, accessed 30 May 2017.

44 For interview with Professor Wang Shuqin, see 'China rates its own citizens – including online behaviour', *Volkskrant*, http://www.volkskrant.nl/buitenland/china-rates-its-own-citizens-including-online-behaviour~a3979668/, 25 April 2015, accessed 4 May 2017.

45 'Planning Outline for the Construction of a Social Credit System', China Copyright and Media, translated by Rogier Creemers, https://chinacopyrightandmedia.wordpress.com/2014/06/14/planning-outline-for-the-construction-of-a-social-credit-system-2014-2020/, 25 April 2015, accessed 3 March 2017.

46 'Open Sesame – Why a Digital "Social Credit" System Makes Sense for China', LinkedIn, https://www.linkedin.com/pulse/open-sesame-why-digital-social-credit-system-makes-sense-majid-%E7%BD%97%E7%B4%A0, 5 November 2015, accessed 30 April 2017.

47 See 'Facebook can recognise you in photos even if you're not looking', *New Scientist*, https://www.newscientist.com/article/dn27761-facebook-can-recognise-you-in-photos-even-if-youre-not-looking/, 22 June 2015, accessed 30 April 2017.

48 See 'OECD Digital Economy Outlook 2015 – Emerging issues: The Internet of Things, OECD Publishing', http://www.keepeek.com/Digital-Asset-Management/oecd/science-and-technology/oecd-digital-economy-outlook-2015/emerging-issues-the-internet-of-things_9789264232440-8-en#page1, 2015, accessed 5 December 2016.

49 See: Zak v Bose Corp, U.S. District Court, Northern District of Illinois, No. 17-02928, https://assets.documentcloud.org/documents/3673948/Zak-v-Bose.pdf, accessed 19 July 2017.

50 'A message to our Bose Connect App customers', Bose, https://www.bose.com.au/en_au/landing_pages/bose_corporation_updates.html, accessed 20 April 2017.
51 See 'We-Vibe's Motion For Approval of Settlement', US District Court, Northern District of Illinois, No. 1:16-cv-08655, https://assets.documentcloud.org/documents/3517061/We-Vibes-Motion-For-Approval-of-Settlement.pdf, accessed 19 July 2017.
52 'Sex toy surveillance: more Wi-Fi enabled devices vulnerable to hacking', *WIRED*, http://www.wired.co.uk/article/we-vibe-sex-toy-surveillance, 5 April 2017, accessed 19 July 2017. Svakom has since said it has addressed the issues and that updated versions of its software were 'completely secure'.
53 See 'Agreement Between TSA and TSA Pre-check Application Expansion', Agenda 21 News, http://agenda21news.com/wp-content/uploads/2015/01/OTA_Articles_for_Pre-check_Application_Expansion.pdf, accessed 5 December 2016.
54 'Password for social media accounts could be required for some to enter country', TechCrunch, https://techcrunch.com/2017/02/08/passwords-for-social-media-accounts-could-be-required-for-some-to-enter-country/, 8 February 2017, accessed 31 May 2017.
55 'A.G. Schneiderman Announces Settlement with Uber to Enhance Rider Privacy', New York State Office of the Attorney General, https://ag.ny.gov/press-release/ag-schneiderman-announces-settlement-uber-enhance-rider-privacy, 6 January 2016, 19 June 2017.
56 See 'RoG Blog': 'Blog.Uber.com/Ridesofglory', *Internet Archive WaybackMachine*, https://web.archive.org/web/20140827195715/http:/blog.uber.com/ridesofglory, 26 March 2012, accessed 19 June 2017.
57 See 'Uber to Pay $20,000 Fine Over "God View" Toll September 2014 Data Breach', TechTimes, http://www.techtimes.com/articles/122410/20160107/uber-to-pay-20-000-fine-over-god-view-tool-september-2014-data-breach.htm, 7 January 2016, accessed 5 December 2016.
58 *The Inevitable*, Kevin Kelly, Viking Press (2016).
59 'Why you should embrace surveillance, not fight it', *WIRED*, https://www.wired.com/2014/03/going-tracked-heres-way-embrace-surveillance/, 3 October 2014, accessed 7 January 2017.

8. In Bots We Trust

1 'Believing in BERT: Using expressive communication to enhance trust and counteract operational error in physical Human-Robot Interaction', Adriana Hamacher, Nadia Bianchi-Berthouze, Anthony Pipe and Kerstin Eder, 2016, 25th IEEE International Symposium on Robot and Human Interactive Communication (RO-MAN), https://arxiv.org/ftp/arxiv/papers/1605/1605.08817.pdf, accessed 12 December 2016.
2 See 'People will lie to robots to avoid "hurting their feelings"', *WIRED*, http://www.wired.co.uk/article/bert-lying-robots-emotions, 23 August 2016, accessed 31 May 2017.
3 See 'Trust Me: Researchers Examine How People and Machines Build Bonds', George Mason University, https://www2.gmu.edu/news/1849, 4 February 2015, accessed 4 March 2017.
4 See 'Can we trust robots?', Mark Coeckelbergh, *Ethics and Information Technology*, http://link.springer.com/article/10.1007/s10676-011-9279-1, 3 September 2011, accessed 4 March 2017.
5 My conversation with Stephen Cave in December 2016 shaped my understanding of the shift from trust in technology being based on it doing something to deciding something.
6 See 'What Is a Robot?', *The Atlantic*, http://www.theatlantic.com/technology/archive/2016/03/what-is-a-human/473166/, 22 March 2015, accessed 5 March 2017.
7 'Computing Machinery and Intelligence', Alan Turing, *Mind*, http://mind.oxfordjournals.org/content/LIX/236/433.full.pdf+html and Oxford University Press (1950).
8 *Speculations Concerning the First Ultraintelligent Machine*, Irving Good, http://www.kushima.org/is/wp-content/uploads/2015/07/Good65ultraintelligent.pdf and Academic Press (1965).
9 See 'Stephen Hawking warns artificial intelligence could end mankind', BBC, http://www.bbc.com/news/technology-30290540, 2 December 2014, accessed 5 March 2017.
10 See 'Computer AI passes Turing test in "world first"', BBC, http://www.bbc.com/news/technology-27762088, 9 June 2014, accessed 5 March 2017.

11 'AI Program beats humans in poker game', BBC News, http://www.bbc.co.uk/news/technology-38812530, 31 January 2017, accessed 4 May 2017.

12 For the price to own a Pepper robot, see 'Pepper the robot's contract bans users from having sex with it', *WIRED*, http://www.wired.co.uk/article/pepper-robot-sex-banned, 23 September 2015, accessed 12 January 2017.

13 See 'No sex please, we're robots! Buyers of hit new "emotional robot" Pepper to sign contract vowing it wont be used indecently', *Daily Mail*, http://www.dailymail.co.uk/news/article-3243051/No-sex-robots-Buyers-hit-new-emotional-robot-Pepper-sign-contract-saying-won-t-used-sex-porno-films.html, 23 September 2015, accessed 5 March 2017.

14 See 'The Great Bot Rush of 2015–2016', Continuations, http://continuations.com/post/135317420600/the-great-bot-rush-of-2015-16, 16 December 2015, accessed 5 March 2017.

15 See 'What Is a Robot?', *The Atlantic*, http://www.theatlantic.com/technology/archive/2016/03/what-is-a-human/473166/, 22 March 2015, accessed 5 March 2017.

16 Ibid.

17 Statistics relating to DoNotPay from Joshua Browder, author interview, 2 March 2017.

18 *Alone Together: Why We Expect More from Technology and Less from Each Other*, Sherry Turkle, Basic Books (2012).

19 'Rise of the Machines: How AI-Driven Personal Assistant Apps Are Shaping Digital Consumer Habits', Verto Analytics, http://research.vertoanalytics.com/hubfs/Files/VertoAnalytics-Personal-Assistant-Report.pdf.

20 MIT Media Lab, 'Making new (robot) friends', https://www.media.mit.edu/posts/making-new-robot-friends, 13 June 2017.

21 Tay gained 50,000 followers and produced nearly 100,000 tweets in less than twenty-four hours after her arrival on Twitter. See 'Why Microsoft's "Tay" AI bot went wrong', TechRepublic, http://www.techrepublic.com/article/why-microsofts-tay-ai-bot-went-wrong/, 24 March 2016, accessed 30 May 2017.

22 See 'Twitter taught Microsoft's AI chatbot to be a racist asshole in less than a day', The Verge, http://www.theverge.com/2016/3/24/11297050/tay-microsoft-chatbot-racist, 24 March 2016, accessed 20 May 2017.

23 Andrew Karpathy explains how neural networks attempt predictive text. See 'Multi-layer Recurrent Neural Networks (LSTM, GRU, RNN) for character-level language models in Torch', Github, https://github.com/karpathy/char-rnn, accessed 9 December 2016.

24 'Tay, Microsoft's AI chatbot, gets a crash course in racism from Twitter', *Guardian*, https://www.theguardian.com/technology/2016/mar/24/tay-microsofts-ai-chatbot-gets-a-crash-course-in-racism-from-twitter, 24 March 2016, accessed 20 May 2017.

25 'We are developing the psyche of software that will sit at the heart of virtual and animated systems' and subsequent quotes from Mark Stephen Meadows, author interview, 1 December 2016.

26 Vidya Narayanan, Philip N. Howard, Bence Kollyani, Mona Elswah, 'Russian Involvement and Junk News During Brexit', Oxford Internet Institute, 19 December 2017.

27 Yuriy Gorodnichenko, Tho Pham, Oleksandr Talavera, 'Social network, sentiment and political outcomes: Evidence from #Brexit', https://editorialexpress.com/cgi-bin/conference/download.cgi?db_name=RESConf2017&paper_id=607, 25 January 2018.

28 See 'The Bot Politic', *New Yorker*, http://www.newyorker.com/tech/elements/the-bot-politic, 31 December 2016, accessed 31 May 2017.

29 See 'Public Predictions for the Future of Workforce Automation', Pew Research Center, http://www.pewinternet.org/2016/03/10/public-predictions-for-the-future-of-workforce-automation/, 10 March 2016, accessed 4 May 2017.

30 See 'The Future of Employment: How Susceptible are Jobs to Computerisation?', Carl Benedikt Frey and Michael Osborne, Oxford Martin School, http://www.oxfordmartin.ox.ac.uk/downloads/academic/The_Future_of_Employment.pdf, 2013, accessed 4 March 2017.

31 See 'An Uncanny Mind, Masahiro Mori on the Uncanny Valley and Beyond', IEEE Spectrum, http://spectrum.ieee.org/automaton/ro

botics/humanoids/an-uncanny-mind-masahiro-mori-on-the-uncanny-valley, 12 June 2012, accessed 4 March 2017.

32 'Now you're talking: human-like robot may one day care for dementia patients', Reuters, http://www.reuters.com/article/singapore-humanoid-idUSKCN0W9120, 7 March 2016, accessed 30 May 2017.

33 'Oh dear! Oh dear! I shall be too late!' See *Alice's Adventures in Wonderland*, Lewis Carroll, Macmillan and Co. (1869), p. 1.

34 See 'The mind in the machine: Anthropomorphism increases trust in the autonomous vehicle', Adam Waytz, Joy Heafner and Nicholas Epley, *Journal of Experimental Social Psychology*, http://www.sciencedirect.com/science/article/pii/S0022103114000067, May 2014, accessed 9 December 2016.

35 See 'The Bot Politic', *New Yorker*, http://www.newyorker.com/tech/elements/the-bot-politic, 31 December 2016, accessed 31 May 2017.

36 See 'Hackers can hijack Wi-Fi Hello Barbie to spy on your children', *Guardian*, https://www.theguardian.com/technology/2015/nov/26/hackers-can-hijack-wi-fi-hello-barbie-to-spy-on-your-children, 26 November 2015, accessed 5 December 2016. Note: ToyTalk Mattel have fixed many of the issues security analyst Andrew Blaich raised, such as removing the weaker SSLv3 ciphers from its servers. 'We are aware of the Bluebox Security Report and are working closely with ToyTalk to ensure the safety and security of Hello Barbie,' said Mattel spokesperson Michelle Chidoni.

37 'One of the key questions is how do we assess the trustworthiness of an intelligent machine?' and subsequent quotes from Stephen Cave, author interview, 6 December 2016.

38 See 'Machine Ethics: The Robots Dilemma', *Nature*, http://www.nature.com/news/machine-ethics-the-robot-s-dilemma-1.17881, 1 July 2015, accessed 4 March 2017.

39 This is also known as Bentham's 'fundamental axiom'. See *A Fragment on Government*, Preface, Jeremy Bentham, Cambridge University Press (1988).

40 *The Right and the Good*, William David Ross, Clarendon Press (2002).

41 See *Machine Ethics*, Michael Anderson and Susan Leigh Anderson, Cambridge University Press (2011).

42 'The Ethical Robot', UConn Today, http://today.uconn.edu/2010/11/the-ethical-robot/, 8 November 2010, accessed 12 December 2016.

43 The trolley problem was invented by the philosophers Philippa Foot and Judith Jarvis in the 1960s. See 'The Problem of Abortion and the Doctrine of the Double Effect', Philippa Foot and Judith Jarvis, http://philpapers.org/archive/FOOTPO-2.pdf, from the *Oxford Review*, 1967.

44 'Rise of the machines: are algorithms sprawling out of our control?', *WIRED*, http://www.wired.co.uk/article/technology-regulation-algorithm-control, 1 April 2017.

45 See 'Principles of robotics', Engineering and Physical Sciences Research Council, https://www.epsrc.ac.uk/research/ourportfolio/themes/engineering/activities/principlesofrobotics/, accessed 12 December 2016.

46 See 'The social dilemma of autonomous vehicles', Jean-François Bonnefon, Azim Shariff and Iyad Rahwan, *Science*, 352/6293, https://arxiv.org/abs/1510.03346, 4 July 2016, accessed 12 December 2016.

47 See 'Driverless cars are colliding with the creepy Trolley Problem', *Washington Post*, https://www.washingtonpost.com/news/innovations/wp/2015/12/29/will-self-driving-cars-ever-solve-the-famous-and-creepy-trolley-problem/?utm_term=.44210b7e6797, 29 December 2015, accessed 12 December 2016.

48 See 'Why robots need to be able to say "no"', The Conversation, http://www.wired.co.uk/article/technology-regulation-algorithm-control, 8 April 2016, accessed 12 December 2016.

9. Blockchain Part I: The Digital Gold Rush

1 For William H. Furness's account of his experience in Yap, see *The Island of Stone Money: Uap of the Carolines*, William H. Furness, J. B. Lippincott Co. (1910), pp. 94–106.

2 See 'David O'Keefe: The King of Hard Currency', *Smithsonian Magazine*, http://www.smithsonianmag.com/history/david-okeefe-the-king-of-hard-currency-37051930/, 28 July 2011, accessed 2 March 2017.

3 See 'When Bitcoin Grows Up', *London Review of Books*, http://www.lrb.co.uk/v38/no8/john-lanchester/when-bitcoin-grows-up?utm_content=bufferc8f7f&utm_medium=social&utm_source=twitter.com&utm_campaign=buffer, 21 April 2016, accessed 2 March 2017.

4 See *The Island of Stone Money: Uap of the Carolines*, William H. Furness, J. B. Lippincott Co. (1910), p. 98.

5 See 'The Island of Stone Money', Milton Friedman, http://www.karlwhelan.com/IMB/Friedman-Yap.pdf, Stanford University, February 1991.

6 For the estimated total amount of money in the world in terms of value, see *The Doctor*, Dr Karl Kruszelnicki, Macmillan Australia (2016), p. 132.

7 *Double Entry*, Jane Gleeson-White, Allen & Unwin (2011).

8 See 'Bitcoin: A Peer-to-Peer Electronic Cash System', Satoshi Nakamoto, https://bitcoin.org/bitcoin.pdf, 24 May 2009, accessed 2 March 2017.

9 See 'Bitcoin open source implementation of P2P currency', Satoshi Nakamoto, P2P foundation, http://p2pfoundation.ning.com/forum/topics/bitcoin-open-source, 11 February 2009, accessed 2 March 2017.

10 See 'Disruptions: Betting on a Coin with no realm', *New York Times*, https://bits.blogs.nytimes.com/2013/12/22/disruptions-betting-on-bitcoin/?_r=0&mtrref=www.forbes.com&gwh=EF12B4F946D4DF3A60818B678EA05D1B&gwt=pay, 22 December 2013, accessed 2 March 2017.

11 For the bitcoin price surge in January 2017, see 'Bitcoin Price Soars, Fueled by Speculation and Global Currency Turmoil', *New York Times*, https://www.nytimes.com/2017/01/03/business/dealbook/bitcoin-price-soars-fueled-by-speculation-and-global-currency-turmoil.html, 3 January 2017, accessed 2 March 2017.

12 Wei Dai, 'b-money', http://www.weidai.com/bmoney.txt, 1998, accessed 2 March 2017.

13 'Back to the Future: Adam Back Remembers the Cypherpunk Revolution and the Origins of Bitcoin', *Bitcoin Magazine*, https://bitcoinmagazine.com/articles/back-future-adam-back-remembers-cypherpunk-revolution-origins-bitcoin-1441741053/, 8 September 2015, accessed 2 March 2017.

14 For the number of bitcoin transactions, see https://blockchain.info/charts/n-transactions-total, accessed 25 June 2018.

15 For the number of bitcoin nodes, see 'Nodes', Bitnodes, https://bitnodes.21.co/dashboard/?days=90, accessed 14 February 2017.

16 See 'The Satoshi Affair', *London Review of Books*, https://www.lrb.co.uk/v38/n13/andrew-ohagan/the-satoshi-affair, 30 June 2016, accessed 14 February 2017.

17 See 'Engineering the Bitcoin Gold Rush: An Interview with Yifu Guo, Creator of the First Purpose-Built Miner', *Motherboard*, https://motherboard.vice.com/en_us/article/engineering-the-bitcoin-gold-rush-an-interview-with-yifu-guo-creator-of-the-first-asic-based-miner, 27 March 2013, accessed 2 March 2017.

18 See 'In the beginning – Trusted Disrupted: Bitcoin and the Blockchain' (Episode 1), https://techcrunch.com/2016/10/10/watch-the-first-episode-of-our-new-series-trust-disrupted-bitcoin-and-the-blockchain/, 10 October 2016, accessed 2 March 2017.

19 See 'Total Number of Transactions', *Blockchain*, https://blockchain.info/charts/n-transactions-total, 15 March 2018.

20 See *The Age of Cryptocurrency: How Bitcoin and the Blockchain are Challenging the Global Economic Order*, Paul Vigna and Michael J. Casey, Picador (2016).

21 'Sichuan – a Paradise of Food and Modern Agriculture', HKTDC, http://www.hktdc.com/web/featured_suppliers/sichuan/index.html, accessed 31 May 2017.

22 For the number of dams on the Min Jiang, see 'Mapping China's "Dam Rush"', Wilson Center, https://www.wilsoncenter.org/publication/interactive-mapping-chinas-dam-rush, 21 March 2014, accessed 2 March 2017.

23 Zhu Rei is interviewed in Episode 2 of the TechCrunch series on Blockchain. See 'Trusted Disrupted: Bitcoin and the Blockchain', TechCrunch, https://techcrunch.com/2016/10/10/watch-trust-disrupted-bitcoin-and-the-blockchain-episode-two/, 10 October 2016, accessed 2 March 2017.

24 See 'The magic of mining', *The Economist*, http://www.economist.com/news/business/21638124-minting-digital-currency-has-become-

big-ruthlessly-competitive-business-magic, 8 January 2015, accessed 2 March 2017.

25 See 'How China took center stage in Bitcoin's civil war', *New York Times*, https://www.nytimes.com/2016/07/03/business/dealbook/bitcoin-china.html, 29 June 2016, accessed 2 March 2017.

26 See 'Bitcoin open source implementation of P2P currency', Satoshi Nakamoto, P2P foundation, http://p2pfoundation.ning.com/forum/topics/bitcoin-open-source, 11 February 2009, accessed 2 March 2017.

27 See 'Missing: hard drive containing Bitcoins worth £4m in Newport landfill site', *Guardian*, https://www.theguardian.com/technology/2013/nov/27/hard-drive-bitcoin-landfill-site, 28 November 2013, accessed 2 March 2017.

28 For the list of countries that banned bitcoin, see 'Top 10 Countries in Which Bitcoin is Banned', CryptoCoinsNews, https://www.cryptocoinsnews.com/top-10-countries-bitcoin-banned/, 27 May 2015, accessed 2 March 2017.

29 Larry Summers interviewed in Episode 1 of the TechCrunch series on Blockchain, see 'In the beginning – Trusted Disrupted: Bitcoin and the Blockchain', TechCrunch,https://techcrunch.com/2016/10/10/watch-the-first-episode-of-our-new-series-trust-disrupted-bitcoin-and-the-blockchain/, 10 October 2016, accessed 5 March 2017.

30 See 'The great chain of being sure about things', *The Economist*, http://www.economist.com/news/briefing/21677228-technology-behind-bitcoin-lets-people-who-do-not-know-or-trust-each-other-build-dependable, 31 October 2015, accessed 30 May 2017.

10. *Blockchain Part II: The Truth Machine*

1 'The Dao Attacked: Code Issue Leads to $60 Million Ether Theft', CoinDesk, http://www.coindesk.com/dao-attacked-code-issue-leads-60-million-ether-theft/, 17 June 2016, accessed 1 March 2017.

2 See 'Ether Price Plummets; Ethereum DAO May Be Hacked', CryptoCoinNews, https://www.cryptocoinsnews.com/ether-price-plumets-ethereum-dao-may-be-hacked, 17 June 2016, accessed 1 March 2017.

3 See 'Understanding the DAO Attack', CoinDesk, http://www.coindesk.com/understanding-dao-hack-journalists/, 25 June 2016, accessed 1 March 2017.

4 Slock.it is a German tech start-up building 'smart locks' that connect all kinds of things – cars, bikes, even a front door – to the blockchain.

5 See 'Why Bitcoin Matters', *New York Times*, https://dealbook.nytimes.com/2014/01/21/why-bitcoin-matters/, 21 January 2014, accessed 1 March 2017.

6 'DAOs, DACs, DAs and More: An Incomplete Terminology Guide', Ethereum Blog, https://blog.ethereum.org/2014/05/06/daos-dacs-das-and-more-an-incomplete-terminology-guide/, 6 May 2014, accessed 4 May 2017.

7 'Can this 22-year old coder out-bitcoin bitcoin?', *Fortune*, http://fortune.com/ethereum-blockchain-vitalik-buterin/, 27 September 2016, accessed 1 March 2017.

8 'The Uncanny Mind that Built Ethereum', Backchannel, https://backchannel.com/the-uncanny-mind-that-built-ethereum-9b448dc9d14f#.wmpr48it1, 13 June 2016, accessed 1 March 2017.

9 'Ethereum', Ethereum, https://www.ethereum.org/foundation, accessed 1 March 2017.

10 'Bootstrapping a Decentralized Autonomous Corporation: Part 1', *Bitcoin Magazine*, https://bitcoinmagazine.com/articles/bootstrapping-a-decentralized-autonomous-corporation-part-i-1379644274/, 19 September 2013, accessed 1 March 2017.

11 'The Uncanny Mind that Built Ethereum', Backchannel, https://backchannel.com/the-uncanny-mind-that-built-ethereum-9b448dc9d14f#.wmpr48it1, 13 June 2016, accessed 1 March 1 2017.

12 Some autonomous companies exist now like Weifund, which is decentralized crowdfunding without an intermediary like Kickstarter. See http://weifund.io/. Other examples: http://www.gdi.ch/en/Think-Tank/GDI-Trend-News/News-Detail/Uber-without-Uber-Platform-cooperativism-as-the-new-sharing-economy.

13 For the Ethereum White Paper, See 'White Paper', Github, https://github.com/ethereum/wiki/wiki/White-Paper, accessed 1 March 2017.

14 'The great chain of being sure about things', *The Economist*, http://www.economist.com/news/briefing/21677228-technology-behind-bitcoin-lets-people-who-do-not-know-or-trust-each-other-build-dependable, 31 October 2015, accessed 30 May 2017.

15 Ibid.

16 In April 2016, Bob Sauchelli purchased his first excess energy, 195 credits for $0.07 each, directly from his neighbour, Eric Frumin, via Transactive Grid. No energy company required. See 'Ethereum Used for "First" Paid Energy Trade Using Blockchain Tech', CoinDesk, http://www.coindesk.com/ethereum-used-first-paid-energy-trade-using-blockchain-technology, 11 April 2016, accessed 1 March 1, 2017.

17 See Tom Standage on Babbage Podcast from *The Economist* interview with Buterin: 'Vitalek Buterin on his long term goals for Ethereum', Bitcuners, http://blog.bitcuners.org/post/143849632438/vitalik-buterin-on-his-long-term-goals-for, 4 May 2016, accessed 1 March 2017.

18 'Havocscope Black Market', Havocscope, http://www.havocscope.com/, accessed 1 March 2017.

19 'A $50 million hack just showed that the DAO was all too human', *WIRED*, https://www.wired.com/2016/06/50-million-hack-just-showed-dao-human/, 18 June 2016, accessed 1 March 2017.

20 See 'An Open letter', Pastebin, http://pastebin.com/CcGUBgDG, 18 June 2016, accessed 23 February 2017.

21 'Can this 22-year-old coder out-bitcoin bitcoin', *Fortune*, http://fortune.com/ethereum-blockchain-vitalik-buterin/, 27 September 2016, accessed 31 May 2017.

22 'A $50 million hack just showed that the DAO was all too human', *WIRED*, https://www.wired.com/2016/06/50-million-hack-just-showed-dao-human/, 18 June 2016, accessed 1 March 2017.

23 'The Ethereum Hard Fork is Done', Futurism, https://futurism.com/the-ethereum-hard-fork-is-done/, 20 July 2016, accessed 12 December 2016.

24 This is a great reference for the sequence of events around the DAO attack – see 'Understanding the DAO Attack', CoinDesk, http://www.coindesk.com/understanding-dao-hack-journalists/, 25 June 2016, accessed 12 December 2016.

25 See 'A $50 million hack just showed that the DAO was all too human', *WIRED*, https://www.wired.com/2016/06/50-million-hack-just-showed-dao-human/, 18 June 2016, accessed 1 March 2017.

26 *Mastering Bitcoin: Unlocking Digital Cyrptocurrencies*, Andreas Antonopoulos, O'Reilly Media (2014).

27 See 'Reid Hoffman: Why the blockchain matters', *WIRED*, http://www.wired.co.uk/article/bitcoin-reid-hoffman, 15 May 2015.

28 For Danny Hillis's interview, See 'Disney's Wizards', *Newsweek*, http://europe.newsweek.com/disneys-wizards-172346?rm=eu, 8 November 1997, accessed 1 March 2017.

29 For the number of DApps since first launched, see Ethereum, http://dapps.ethercasts.com/, accessed 10 May 2017.

30 For the number of apps in the Appstore, see https://www.theverge.com/2018/4/5/17204074/apple-number-app-store-record-low-2017-developers-ios, accessed 25 June 2018.

31 'PwC Expert: $1.4 Billion Invested in Blockchain in 2016', CryptoCoins News, https://www.cryptocoinsnews.com/pwc-expert-1-4-billion-invested-blockchain-2016/, 9 November 2016, accessed 1 March 2017.

32 See 'Introducing R3 Corda: A Distributed Ledger Designed for Financial Services', R3 blog, http://www.r3cev.com/blog/2016/4/4/introducing-r3-corda-a-distributed-ledger-designed-for-financial-services, 5 April 2016, accessed 31 May 2017.

33 'It means the history of something, where it came from and where did it go' and subsequent quotes from Leanne Kemp, author interview, 2 March 2017.

34 See 'Crime pays when provenance is broken', Everledger, https://www.everledger.io/, accessed 1 March 2017.

35 For estimates of stolen art, see 'Nazi loot case: much art still untraced – expert', BBC News, http://www.bbc.com/news/world-europe-24801935, 4 November 2013, accessed 1 March 2017.

36 'Nazi Art Loot Returned', *New York Times*, https://www.nytimes.com/2016/07/16/arts/design/nazi-art-loot-returned-to-nazis.html?_r=0, 15 July 2016, accessed 1 March 2017.

37 See American Alliance of Museums website, 'Standards Regarding the Unlawful Appropriation of Objects During the Nazi Era', http://www.aam-us.org/resources/ethics-standards-and-best-practices/collections-stewardship/objects-during-the-nazi-era, accessed 1 March 2017.

38 'Every product has a story' and subsequent quotes from Jessi Baker, author interview, 9 August 2016.

39 See 'Walmart and IBM are Partnering to Put Chinese Pork on a Blockchain', *Fortune*, http://fortune.com/2016/10/19/walmart-ibm-blockchain-china-pork/, 19 October 2016, accessed 1 March 2017.

40 'World Economic Forum Annual Meeting', World Economic Forum, https://www.weforum.org/events/world-economic-forum-annual-meeting-2017, accessed 16 May 2017.

41 See 'Blockchain technology: Redefining trust for a global, digital economy', World Bank, http://blogs.worldbank.org/ic4d/blockchain-technology-redefining-trust-global-digital-economy, 16 June 2016, accessed 1 March 2017.

42 See 'An interview with Hernando de Soto', McKinsey, http://www.mckinsey.com/industries/public-sector/our-insights/an-interview-with-hernando-de-soto, October 2012, accessed 1 March 2017.

43 *The Mystery of Capital: Why Capitalism Triumphs in the West and Fails Everywhere Else*, Hernando de Soto, Basic Books (2000).

44 See *Blockchain Revolution*, Don Tapscott and Alex Tapscott, Portfolio Penguin (2016).

45 See *Land Policies for Growth and Poverty Reduction*, World Bank Policy Research Report, 2003.

46 'Distributed Ledger Technology: Beyond Blockchain', UK Government, https://www.gov.uk/government/uploads/system/uploads/attachment_data/file/492972/gs-16-1-distributed-ledger-technology.pdf, 19 January 2016, accessed 1 March 2017.

47 See 'Bitcoin: A Peer-to-Peer Electronic Cash System', https://bitcoin.org/bitcoin.pdf, accessed 16 May 2017.

48 See 'Why Bitcoin is and isn't like the internet', Joi Ito blog, https://joi.ito.com/weblog/2015/01/23/why-bitcoin-is-.html, 23 January 2015, accessed 1 March 2017.

49 For a timeline of Blythe Masters's work life, see 'Outsmarted high finance vs. human nature', *New Yorker*, http://www.newyorker.com/magazine/2009/06/01/outsmarted, 1 June 2009, accessed 16 May 2017.

50 *Digital Gold: The Untold Story of Bitcoin*, Nathaniel Popper, Harper (2016).

51 See 'The Fintech 2.0 Paper: rebooting financial services', Finextra, https://www.finextra.com/finextra-downloads/newsdocs/the%20fintech%202%200%20paper.pdf, 2015, accessed 10 May 2017.

52 See 'Blythe Masters Tells Banks the Blockchain Changes Everything', Bloomberg, https://www.bloomberg.com/news/features/2015-09-01/blythe-masters-tells-banks-the-blockchain-changes-everything, 1 September 2015, accessed 10 May 2017.

53 'JP Morgan to pay $410 million to settle power market case', Reuters, http://www.reuters.com/article/us-jpmorgan-ferc-idUSBRE96T0NA20130730, 30 July 2013, accessed 10 May 2017.

54 See Don and Alex Tapscott for an explanation of how the blockchain could be an ally or threat for the financial industry, *Blockchain Revolution: How the Technology Behind Bitcoin is Changing Money, Business and the World*, Don Tapscott and Alex Tapscott, Portfolio Penguin (2016).

55 'Goldman Sachs Files Patent for Virtual Settlement Currency', *Financial Times*, https://www.ft.com/content/b0d8f614-997c-11e5-9228-87e603d47bdc, 3 December 2015, accessed 1 March 2017.

56 For the number of banks that have a stake in R3CEV, see 'R3 Home', R3, http://r3members.com/, accessed 1 March 2017.

57 The original five Bitcoin Core developers were Gavin Andresen, Wladimir J. van der Laan, Pieter Wuille, Greg Maxwell and Jeff Garzik. Gavin convinced Mike Hearn to work on Bitcoin Core, see 'Benevolent dictators and disenchanted believers: bitcoin core developers revisited', CoinFox, http://www.coinfox.info/news/reviews/5312-benevolent-dictators-and-disenchanted-believers-bitcoin-core-developers-revisited, 15 April 2016, accessed 1 March 2017.

58 See 'The resolution of the Bitcoin experiment', https://blog.plan99.net/the-resolution-of-the-bitcoin-experiment-dabb30201f7#.idmijyl38, 14 January 2016, accessed 1 March 2017.

59 See 1993 CBC interview with John Allen, 'Tech time warp: 20 years ago, we thought the internet would bring out our nice sides', *WIRED*, https://www.wired.com/2014/06/tech-time-warp-cyber-bullies/, 13 June 2014, accessed 1 March 2017.

Conclusion

1 See 'In No One We Trust', Joseph Stiglitz, *New York Times*, https://opinionator.blogs.nytimes.com/2013/12/21/in-no-one-we-trust/?_r=3, 21 December 2013, accessed 16 May 2017.

2 Eric's story is retold by Shivani Siroya and the Tala team through Skype and email correspondence, 18 January 2017 and 25 February 2017.

3 'We would say to her, "Mom you work so hard and your patients never pay you"' and other quotes from Shivani Siroya, author interview, 18 January 2017.

4 See 'A smart loan for people with no credit history (yet)', TED talk, https://www.ted.com/talks/shivani_siroya_a_smart_loan_for_people_with_no_credit_history_yet/transcript?language=en#t-102209, April 2016, accessed 16 May 2017.

5 See 'Do Your Research Before Changing the World', Collaborative Fund, http://www.collaborativefund.com/blog/shivani-siroya/, 8 March 2016, accessed 16 May 2017.

6 'This is data that would not be found on a paper trail or in any formal financial record': I have combined quotes from my interview with her and her TED speech where she makes a similar point.

7 See 'An Implosion of Trust', Edelman, http://www.edelman.com.au/magazine/posts/an-implosion-of-trust/, accessed 18 January 2017.

8 See 'Uber CEO on Driver "Assault": It's Not Real and We're Not Responsible', Valleywag, http://valleywag.gawker.com/uber-ceo-on-driver-assault-its-not-real-and-were-n-1323533057, 16 September 2013, accessed 16 May 2017.

9 See 'Uber Chief's Email to Employees', *New York Times*, https://www.nytimes.com/interactive/2017/02/03/technology/document-Kalanick-email.html, 3 February 2017, accessed 31 May 2017.

10 'The Second Coming', *The Collected Poems of W. B. Yeats*, Wordsworth Editions (1994).

11 For a description of the current state of ceaseless adaption and change, see *The Inevitable*, Kevin Kelly, Viking Press (2016).

12 There is a beautiful post by the Iranian blogger Hossein Derakhshan, who was imprisoned for six years for his online activity, who talks about this point. See 'The Web We Have to Save', Medium, https://medium.com/matter/the-web-we-have-to-save-2eb1fe15a426#, 14 July 2015, accessed 16 May 2017.

Further Reading

Academic Papers and Research

Aldridge, J., and Décary-Hétu, D., 'Hidden wholesale: The drug diffusing capacity of online drug cryptomarkets', *International Journal of Drug Policy*, 35, 7–15 (2016), https://doi.org/10.1016/j.drugpo.2016.04.020.

Ashraf, N., Bohnet, I., and Piankov, N., 'Decomposing trust and trustworthiness', *Experimental Economics*, 9, 193–208 (2006), https://doi.org/10.1007/s10683-006-9122-4.

Ashraf, N., Camerer, C. F., and Loewenstein, G., 'Adam Smith, Behavioral Economist', *Journal of Economic Perspectives*, 19(3), 131–45 (2005), retrieved from https://www.cmu.edu/dietrich/sds/docs/loewenstein/AdamSmith.pdf.

Barr, A., 'Familiarity and Trust: An experimental investigation', WPS99-23, Centre for the Study of African Economies, Department of Economics, University of Oxford (1999), retrieved from http://www.csae.ox.ac.uk/materials/papers/9923text.PDF.

Boyd, Danah, Levy, Karen, Marwick, A., 'The Networked Nature of Algorithmic Discrimination', Open Technology Institute, 53–7 (October 2014), retrieved from https://www.danah.org/papers/2014/DataDiscrimination.pdf.

Castaldo, Sandro, Premazzi, Katia, Zerbini, Fabrizio, 'The Meaning(s) of Trust. A Content Analysis on the Diverse Conceptualizations of Trust in Scholarly Research on Business Relationships', *Journal of Business Ethics*, 96(4), 657–68 (2010), https://doi.org/10.1007/s10551-010-0491-4.

Chandrasekhar, A. G., Kinnan, C., Larreguy, H., 'Social Networks As Contract Enforcement: Evidence from a Lab Experiment in the Field',

The National Bureau of Economic Research, 1–48 (June 2014). Retrieved from https://web.stanford.edu/~arungc/CKL.pdf.

Cheshire, C., 'Online Trust, Trustworthiness, or Assurance?', *Daedalus*, 140(4), 49–58 (2011), https://doi.org/10.1162/DAED_a_00114.

Cheshire, C., Antin, J., Cook, K. S., Churchill, E., 'General and Familiar Trust in Websites', *Knowledge, Technology & Policy*, 23, 311–31 (2010), https://doi.org/10.1007/s12130-010-9116-6.

Cheshire, C., and Cook, K. S., 'The Emergence of Trust Networks Under Uncertainty – Implications for Internet Interactions', *Analyse & Kritik*, 26, 220–40 (2004), retrieved from http://www.analyse-und-kritik.net/Dateien/56c1dc81bfc40_ak_cheshire_2004.pdf.

Christin, N., 'Traveling the Silk Road: A measurement analysis of a large anonymous online marketplace', Carnegie Mellon INI, retrieved from https://www.andrew.cmu.edu/user/nicolasc/publications/TR-CMU-CyLab-12-018.pdf.

Cook, Karen S., Cheshire, Coye, Gerbasi, Alexandra, and Aven, Brandy, 'Assessing Trustworthiness in Providers of Online Goods and Services' in Karen S. Cook, Chris Snijders, Vincent Buskins, Coye Cheshire (eds), *eTrust: Forming Relationships in the Online World*, pp. 189–214, Russell Sage Foundation (2009), retrieved from https://www.ischool.berkeley.edu/research/publications/2009/assessing-trustworthiness-providers-online-goods-and-services.

Darling, K., '"Who's Johnny?" Anthropomorphic Framing in Human-Robot Interaction, Integration, and Policy', in R. Jenkins, P. Lin, G. Bekey, K. Abney (eds), *ROBOT ETHICS 2.0*, pp. 173–88, Oxford University Press (2017), https://doi.org/10.2139/ssrn.2588669.

Das, T. K., and Teng, B.-S., 'The Risk-Based View of Trust: A Conceptual Framework', *Journal of Business and Psychology*, 19(1), 85–116 (2004), https://doi.org/10.1023/B:JOBU.0000040274.23551.1b.

Deutsch, M., 'Trust and suspicion', *Journal of Conflict Resolution*, 2(4), 265–79 (1958), https://doi.org/10.1177/002200275800200401.

Erickson, T., and Kellogg, W. A., 'Social Translucence: An Approach to Designing Systems that Support Social Processes', *ACM Transactions on Computer-Human Interaction*, 7(1), 59–83 (2000), retrieved from http://

citeseerx.ist.psu.edu/viewdoc/download?doi=10.1.1.89.989&rep=rep1&type=pdf.

Evans, A. M., and Krueger, J. I., 'The Psychology (and Economics) of Trust', *Social and Personality Psychology Compass*, 3(6), 1003–17 (2009), https://doi.org/10.1111/j.1751-9004.2009.00232.x.

Evensky, J. M., 'Adam Smith's Essentials: On Trust, Faith, and Free Markets', *Journal of the History of Economic Thought*, 33(2), (2011), retrieved from http://surface.syr.edu/ecn/6.

Falvello, V., Vinson, M., Ferrari, C., and Todorov, A., 'The Robustness of Learning about the Trustworthiness of Other People', *Social Cognition*, 33(5), 368–86 (2015), https://doi.org/10.1521/soco.2015.33.5.368.

Fisher, C., 'The trouble with "trust" in news media', *Communication Research and Practice*, 2(4), 451–65 (2016), https://doi.org/10.1080/22041451.2016.1261251.

Guinnane, T. W., 'Trust: A Concept Too Many', Economic Growth Center, Yale University (2005), retrieved from http://www.econ.yale.edu/~egcenter/.

Hovland, C. I., chapter 4 in *Communication and Persuasion: Psychological Studies of Opinion Change* by Carl I. Hovland, Irving L. Janis and Harold H. Kelley, New Haven: Yale University Press (1953), retrieved from https://trove.nla.gov.au/work/10553521.

Jessen, J., and Jørgensen, A. H., 'Aggregated trustworthiness: Redefining online credibility through social validation', *First Monday*, 17(1) (2012), https://doi.org/10.5210/fm.v17i1.3731.

Johnson-George, C., and Swap, W. C., 'Measurement of specific interpersonal trust: Construction and validation of a scale to assess trust in a specific other', *Journal of Personality and Social Psychology*, 43(6), 1306–17 (1982), https://doi.org/10.1037/0022-3514.43.6.1306.

Kiyonari, T., Yamagishi, T., Cook, K. S., and Cheshire, C., 'Does Trust Beget Trustworthiness? Trust and Trustworthiness in Two Games and Two Cultures: A Research Note', *Social Psychology Quarterly*, 69, 270–83 (2006), http://journals.sagepub.com/doi/abs/10.1177/019027250606900304.

Larzelere, R. E., and Huston, T. L., 'The Dyadic Trust Scale: Toward Understanding Interpersonal Trust in Close Relationships', *Journal of*

Marriage and Family, 42(3), 595–604 (1980), https://doi.org/10.2307/351903.

Möllering, G., 'Trust Beyond Risk: The Leap Of Faith', chapter 5 in *Trust: Reason, Routine, Reflexivity*, Oxford: Elsevier (2006), retrieved from https://www.kent.ac.uk/scarr/events/Mollering.pdf.

Parigi, P., State, B., Dakhlallah, D., Corten, R., and Cook, K., 'A Community of Strangers: The Dis-Embedding of Social Ties', *PLOS One*, 8(7), e67388 (2013), https://doi.org/10.1371/journal.pone.0067388.

Porter, Stephen, England, Laura, Juodis, Marcus, ten Brinke, Leanne, and Wilson, K., 'Is the face a window to the soul? Investigation of the accuracy of intuitive judgments of the trustworthiness of human faces', *Canadian Journal of Behavioural Science*, 40(3), 171–7 (2008), retrieved from http://psycnet.apa.org/buy/2008-09193-005.

Rousseau, D. M., Sitkin, S. B., Burt, R. S., and Camerer, C., 'Not so different after all: A cross-discipline view of trust', *Academy of Management Review*, 23(3), 393–404 (1998), https://www.researchgate.net/publication/50313187_Not_So_Different_After_All_A_Cross-discipline_View_of_Trust.

Shockley, E., Neal, T. M. S., Pytlikzillig, L. M., and Bornstein, B. H. (eds), *Interdisciplinary Perspectives on Trust: Towards Theoretical and Methodological Integration* (2016), https://doi.org/10.1007/978-3-319-22261-5.

Stanford University Center for the Study of Language and Information (US), 'Trust', *Stanford Encyclopedia of Philosophy*, https://plato.stanford.edu/entries/trust/.

Ten Brinke, L., and Adams, G. S., 'Saving face? When emotion displays during public apologies mitigate damage to organizational performance', *Organizational Behavior and Human Decision Processes*, 130, 1–12 (2015), https://doi.org/10.1016/J.OBHDP.2015.05.003.

Tzanetakis, M., Kamphausen, G., Werse, B., and von Laufenberg, R., 'The transparency paradox. Building trust, resolving disputes and optimising logistics on conventional and online drugs markets', *International Journal of Drug Policy*, 35, 58–68 (2016), https://doi.org/10.1016/j.drugpo.2015.12.010.

Watson, M. L., 'Can There Be Just One Trust? A Cross-Disciplinary Identification Of Trust Definitions And Measurement', University of Miami (2005), https://instituteforpr.org/wp-content/uploads/2004_Watson.pdf.

Zand, D. E., 'Trust and Managerial Problem Solving', *Administrative Science Quarterly*, 17(2), 229 (1972), https://doi.org/10.2307/2393957.

Zhang, J., Ghorbani, A. A., and Cohen, R., 'A familiarity-based trust model for effective selection of sellers in multiagent e-commerce systems', *International Journal of Information Security*, 6(5), 333–44 (2007), https://doi.org/10.1007/s10207-007-0025-y.

Recommended Reading

Ariely, D., *Predictably Irrational: The Hidden Forces That Shape Our Decisions*, HarperCollins Publishers Ltd (2009).

Axelrod, R., *The Evolution of Cooperation*, Basic Books (1984).

Bachmann, R., and Zaheer, A., (eds), *The Handbook of Trust Research*, Edward Elgar Publishing Limited (2006).

Bartlett, J., *The Dark Net: Inside the Digital Underworld*, Windmill Books (2015).

Bostrom, N., *Superintelligence: Paths, Dangers, Strategies*, Oxford University Press (2014).

Brooks, D., *The Social Animal: The Hidden Sources of Love, Character, and Achievement*, Short Books (2011).

Coleman, J. S., *Foundations of Social Theory*, Belnap Press (1994).

DeSteno, D., *The Truth About Trust: How it Determines Success in Life, Love, Learning, and More*, Plume (2015).

Ferguson, N., *The Great Degeneration: How Institutions Decay and Economies Die*, Penguin Press (2013).

Floridi, L., *The 4th Revolution: How the Infosphere is Reshaping Human Reality*, Oxford University Press (2016).

Fukuyama, F., *Trust: The Social Virtues and the Creation of Prosperity*, Free Press (1995).

Gallagher, L., *The Airbnb Story: How Three Ordinary Guys Disrupted an Industry, Made Billions . . . and Created Plenty of Controversy*, Houghton Mifflin Harcourt (2017).

Harari, Y. N., *Sapiens: A Brief History of Humankind*, Harvill Secker (2011).

Hardin, R., *Trust and Trustworthiness*, The Russel Sage Foundation (2002).

Hayes, C., *Twilight of the Elites: America after Meritocracy*, Broadway Books (2013).

Jansson, F., *Cryptocurrency: The Fundamental Guide to Trading, Investing, and Mining in Blockchain with Bitcoin and More*, CreateSpace Independent Publishing Platform (2017).

Jones, O., *The Establishment and How They Get Away With It*, Allen Lane (2014).

Kahneman, D., *Thinking, Fast and Slow*, Penguin Books (2012).

Lewis, M., *The Undoing Project: A Friendship that Changed the World*, W. W. Norton & Company (2016).

Nichols, T. M., *The Death of Expertise: The Campaign Against Established Knowledge and Why it Matters*, OUP USA (2017).

Obermayer, B., and Obermaier, F., *The Panama Papers: Breaking the Story of How the Rich and Powerful Hide their Money*, Oneworld Publications (2016).

O'Neil, C., *Weapons of Math Destruction: How Big Data Increases Inequality and Threatens Democracy*, Crown (2016).

O'Neill, O., *A Question of Trust*, Cambridge University Press (2002).

Popper, N., *Digital Gold: Bitcoin and the Inside Story of the Misfits and Millionaires Trying to Reinvent Money*, HarperCollins (2015).

Porter, E., *Alibaba's World: How a Remarkable Chinese Company is Changing the Face of Global Business*, St Martin's Press (2015).

Putnam, R. D., *Bowling Alone: The Collapse and Revival of American Community*, Simon & Schuster (2000).

Schneier, B., *Liars and Outliers: Enabling the Trust that Society Needs to Thrive*, Wiley (2012).

Stone, B., *The Upstarts: How Uber, Airbnb, and the Killer Companies of the New Silicon Valley are Changing the World*, Little, Brown and Company (2017).

Surowiecki, J., *The Wisdom of Crowds: Why the Many are Smarter than the Few and How Collective Wisdom Shapes Business, Economies, Societies and Nations*, Anchor (2005).

Tapscott, A., and Tapscott, D., *Blockchain Revolution: How the Technology Behind Bitcoin is Changing Money, Business and the World*, Penguin Portfolio (2016).

Wallach, W., and Allen, C., *Moral Machines: Teaching Robots Right from Wrong*, Oxford University Press (2010).

Zak, P. J., *Trust Factor: The Science of Creating High-Performance Companies*, AMACOM (2017).

Index